T0298351

An Introduction to
Mathematical Modeling

WILEY SERIES IN COMPUTATIONAL MECHANICS

Series Advisors: **René de Borst, Perumal Nithiarasu, Tayfun Tezduyar, Genki Yagawa, Tarek Zohdi**

Introduction to Finite Element Analysis: Formulation, Verification and Validation
Barna Szabó, Ivo Babuška (March 2011)

An Introduction to Mathematical Modeling: A Course in Mechanics
J. Tinsley Oden (September 2011)

Computational Mechanics of Discontinua
Antonio A Munjiza, Earl Knight, Esteban Rougier (October 2011)

An Introduction to Mathematical Modeling

A Course in Mechanics

J. Tinsley Oden

A JOHN WILEY & SONS, INC., PUBLICATION

For general information on our other products and services please contact our Customer Care Department within the United States at (800) 762-2974, outside the United States at (317) 572-3993 or fax (317) 572-4002.

Wiley also publishes its books in a variety of electronic formats. Some content that appears in print, however, may not be available in electronic formats. For more information about Wiley products, visit our web site at www.wiley.com.

Library of Congress Cataloging-in-Publication Data:

Oden, J. Tinsley (John Tinsley), 1936–
   An introduction to mathematical modeling : a course in mechanics / J. Tinsley Oden.
      p. cm. — (Wiley series in computational mechanics)
   Includes bibliographical references and index.
    ISBN 978-1-118-01903-0 (hardback)
   1. Mechanics, Analytic. I. Title.
   QA807.O34 2011
   530—dc23   2011012204

Printed in the United States of America.

oBook: 978-1-118-10573-3
ePDF: 978-1-118-10576-4
ePub: 978-1-118-10574-0

10 9 8 7 6 5 4 3 2

To Walker and Lee

CONTENTS

Preface   **xiii**

I   Nonlinear Continuum Mechanics   **1**

1   Kinematics of Deformable Bodies   **3**
   1.1   Motion   4
   1.2   Strain and Deformation Tensors   7
   1.3   Rates of Motion 10
   1.4   Rates of Deformation 13
   1.5   The Piola Transformation 15
   1.6   The Polar Decomposition Theorem 19
   1.7   Principal Directions and Invariants of Deformation and
         Strain 20
   1.8   The Reynolds' Transport Theorem 23

2   Mass and Momentum   **25**
   2.1   Local Forms of the Principle of Conservation of Mass . . 26
   2.2   Momentum 28

3   Force and Stress in Deformable Bodies   **29**

4   The Principles of Balance of Linear and Angular Momentum 35
   4.1   Cauchy's Theorem: The Cauchy Stress Tensor 36

4.2   The Equations of Motion (Linear Momentum) 38
4.3   The Equations of Motion Referred to the Reference
      Configuration: The Piola–Kirchhoff Stress Tensors . . . 40
4.4   Power 42

5   The Principle of Conservation of Energy   45
5.1   Energy and the Conservation of Energy 45
5.2   Local Forms of the Principle of Conservation of Energy . 47

6   Thermodynamics of Continua and the Second Law   49

7   Constitutive Equations   53
7.1   Rules and Principles for Constitutive Equations 54
7.2   Principle of Material Frame Indifference 57
      7.2.1   Solids 57
      7.2.2   Fluids 59
7.3   The Coleman–Noll Method: Consistency with the
      Second Law of Thermodynamics 60

8   Examples and Applications   63
8.1   The Navier–Stokes Equations for Incompressible Flow . 63
8.2   Flow of Gases and Compressible Fluids:
      The Compressible Navier–Stokes Equations 66
8.3   Heat Conduction 67
8.4   Theory of Elasticity 69

**II   Electromagnetic Field Theory and Quantum
Mechanics   73**

9   Electromagnetic Waves   75
9.1   Introduction 75
9.2   Electric Fields 75
9.3   Gauss's Law 79
9.4   Electric Potential Energy 80
      9.4.1   Atom Models 80
9.5   Magnetic Fields 81

9.6 Some Properties of Waves 84
9.7 Maxwell's Equations 87
9.8 Electromagnetic Waves 91

10 Introduction to Quantum Mechanics 93
10.1 Introductory Comments 93
10.2 Wave and Particle Mechanics 94
10.3 Heisenberg's Uncertainty Principle 97
10.4 Schrödinger's Equation 99
    10.4.1 The Case of a Free Particle 99
    10.4.2 Superposition in $\mathbb{R}^n$ 101
    10.4.3 Hamiltonian Form 102
    10.4.4 The Case of Potential Energy 102
    10.4.5 Relativistic Quantum Mechanics 102
    10.4.6 General Formulations of Schrödinger's Equation 103
    10.4.7 The Time-Independent Schrödinger Equation . . 104
10.5 Elementary Properties of the Wave Equation 104
    10.5.1 Review 104
    10.5.2 Momentum 106
    10.5.3 Wave Packets and Fourier Transforms 109
10.6 The Wave–Momentum Duality 110
10.7 Appendix: A Brief Review of Probability Densities . . . 111

11 Dynamical Variables and Observables in Quantum
    Mechanics: The Mathematical Formalism 115
11.1 Introductory Remarks 115
11.2 The Hilbert Spaces $L^2(\mathbb{R})$ (or $L^2(\mathbb{R}^d)$) and $H^1(\mathbb{R})$ (or
    $H^1(\mathbb{R}^d)$) 116
11.3 Dynamical Variables and Hermitian Operators 118
11.4 Spectral Theory of Hermitian Operators: The Discrete
    Spectrum 121
11.5 Observables and Statistical Distributions 125
11.6 The Continuous Spectrum 127
11.7 The Generalized Uncertainty Principle for Dynamical
    Variables 128
    11.7.1 Simultaneous Eigenfunctions 130

12 Applications: The Harmonic Oscillator and the Hydrogen Atom   131
12.1 Introductory Remarks 131
12.2 Ground States and Energy Quanta: The Harmonic Oscillator 131
12.3 The Hydrogen Atom 133
    12.3.1 Schrödinger Equation in Spherical Coordinates . 135
    12.3.2 The Radial Equation 136
    12.3.3 The Angular Equation 138
    12.3.4 The Orbitals of the Hydrogen Atom 140
    12.3.5 Spectroscopic States 140

13 Spin and Pauli's Principle   145
13.1 Angular Momentum and Spin 145
13.2 Extrinsic Angular Momentum 147
    13.2.1 The Ladder Property: Raising and Lowering States 149
13.3 Spin 151
13.4 Identical Particles and Pauli's Principle 155
13.5 The Helium Atom 158
13.6 Variational Principle 161

14 Atomic and Molecular Structure   165
14.1 Introduction 165
14.2 Electronic Structure of Atomic Elements 165
14.3 The Periodic Table 169
14.4 Atomic Bonds and Molecules 173
14.5 Examples of Molecular Structures 180

15 Ab Initio Methods: Approximate Methods and Density Functional Theory   189
15.1 Introduction 189
15.2 The Born–Oppenheimer Approximation 190
15.3 The Hartree and the Hartree–Fock Methods 194
    15.3.1 The Hartree Method 196
    15.3.2 The Hartree–Fock Method 196
    15.3.3 The Roothaan Equations 199

15.4 Density Functional Theory 200
    15.4.1 Electron Density 200
    15.4.2 The Hohenberg–Kohn Theorem 205
    15.4.3 The Kohn–Sham Theory 208

III   Statistical Mechanics 213

16 Basic Concepts: Ensembles, Distribution Functions, and Averages 215

16.1 Introductory Remarks 215
16.2 Hamiltonian Mechanics 216
    16.2.1 The Hamiltonian and the Equations of Motion . . 218
16.3 Phase Functions and Time Averages 219
16.4 Ensembles, Ensemble Averages, and Ergodic Systems . . 220
16.5 Statistical Mechanics of Isolated Systems 224
16.6 The Microcanonical Ensemble 228
    16.6.1 Composite Systems 230
16.7 The Canonical Ensemble 234
16.8 The Grand Canonical Ensemble 239
16.9 Appendix: A Brief Account of Molecular Dynamics . . 240
    16.9.1 Newtonian's Equations of Motion 241
    16.9.2 Potential Functions 242
    16.9.3 Numerical Solution of the Dynamical System . . 245

17 Statistical Mechanics Basis of Classical Thermodynamics 249

17.1 Introductory Remarks 249
17.2 Energy and the First Law of Thermodynamics 250
17.3 Statistical Mechanics Interpretation of the Rate of Work in Quasi-Static Processes 251
17.4 Statistical Mechanics Interpretation of the First Law of Thermodynamics 254
    17.4.1 Statistical Interpretation of $\dot{Q}$ 256
17.5 Entropy and the Partition Function 257
17.6 Conjugate Hamiltonians 259
17.7 The Gibbs Relations 261

17.8 Monte Carlo and Metropolis Methods 262
 17.8.1 The Partition Function for a Canonical Ensemble 263
 17.8.2 The Metropolis Method 264
17.9 Kinetic Theory: Boltzmann's Equation of
Nonequilibrium Statistical Mechanics 265
 17.9.1 Boltzmann's Equation 265
 17.9.2 Collision Invariants 268
 17.9.3 The Continuum Mechanics of Compressible
  Fluids and Gases: The Macroscopic Balance
  Laws 269

Exercises   **273**

Bibliography   **317**

Index   **325**

PREFACE

This text was written for a course on *An Introduction to Mathematical Modeling* for students with diverse backgrounds in science, mathematics, and engineering who enter our program in *Computational Science, Engineering, and Mathematics*. It is not, however, a course on just how to construct mathematical models of physical phenomena. It is a course designed to survey the classical mathematical models of subjects forming the foundations of modern science and engineering at a level accessible to students finishing undergraduate degrees or entering graduate programs in computational science. Along the way, I develop through examples how the most successful models in use today arise from basic principles and modern and classical mathematics. Students are expected to be equipped with some knowledge of linear algebra, matrix theory, vector calculus, and introductory partial differential equations, but those without all these prerequisites should be able to fill in some of the gaps by doing the exercises.

I have chosen to call this a textbook on mechanics, since it covers introductions to continuum mechanics, electrodynamics, quantum mechanics, and statistical mechanics. If mechanics is the branch of physics and mathematical science concerned with describing the motion of bodies, including their deformation and temperature changes, under the action of forces, and if one adds to this the study of the propagation of waves and the transformation of energy in physical systems, then the term mechanics does indeed apply to everything that is covered here.

The course is divided into three parts. Part I is a short course on *nonlinear continuum mechanics*; Part II contains a brief account of *electromagnetic wave theory and Maxwell's equations*, along with an introductory account of *quantum mechanics*, pitched at an undergraduate level but aimed at students with a bit more mathematical sophistication than many undergraduates in physics or engineering; and Part III is a brief introduction to *statistical mechanics* of systems, primarily those in thermodynamic equilibrium.

There are many good treatments of the component parts of this work that have contributed to my understanding of these subjects and inspired their treatment here. The books of Gurtin, Ciarlet, and Batra provide excellent accounts of continuum mechanics at an accessible level, and the excellent book of Griffiths on introductory quantum mechanics is a well-crafted text on this subject. The accounts of statistical mechanics laid out in the book of Weiner and the text of McQuarrie, among others, provide good introductions to this subject. I hope that the short excursion into these subjects contained in this book will inspire students to want to learn more about these subjects and will equip them with the tools needed to pursue deeper studies covered in more advanced texts, including some listed in the references.

The evolution of these notes over a period of years benefited from input from several colleagues. I am grateful to Serge Prudhomme, who proofread early versions and made useful suggestions for improvement. I thank Alex Demkov for reading and commenting on Part II. My sincere thanks also go to Albert Romkes, who helped with early drafts, to Ludovic Chamoin, who helped compile and type early drafts of the material on quantum mechanics, and Kris van der Zee, who helped compile a draft of the manuscript and devoted much time to proofreading and helping with exercises. I am also indebted to Pablo Seleson, who made many suggestions that improved presentations in Part II and Part III and who was of invaluable help in putting the final draft together.

J. Tinsley Oden

Austin, Texas
June 2011

Part I

Nonlinear Continuum Mechanics

CHAPTER 1

KINEMATICS OF DEFORMABLE BODIES

Continuum mechanics models the physical universe as a collection of "deformable bodies," a concept that is easily accepted from our everyday experiences with observable phenomena. Deformable bodies occupy regions in three-dimensional Euclidean space $\mathcal{E}$, and a given body will occupy different regions at different times. The subsets of $\mathcal{E}$ occupied by a body $\mathcal{B}$ are called its *configurations*. It is always convenient to identify one configuration in which the geometry and physical state of the body are known and to use that as the *reference configuration*; then other configurations of the body can be characterized by comparing them with the reference configuration (in ways we will make precise later).

For a given body, we will assume that the reference configuration is an open, bounded, connected subset Ω_0 of $\mathbb{R}^3$ with a smooth boundary $\partial\Omega_0$. The body is made up of physical points called *material points*. To identify these points, we assign each a vector X and we identify the components of X as the coordinates of the place occupied by the material point when the body is in its reference configuration relative to a fixed Cartesian coordinate system.

It is thus important to understand that the body $\mathcal{B}$ is a *non-denumerable* set of material points X. This is the fundamental hypoth-

An Introduction to Mathematical Modeling: A Course in Mechanics, First Edition. By J. Tinsley Oden
© 2011 John Wiley & Sons, Inc. Published 2011 by John Wiley & Sons, Inc.

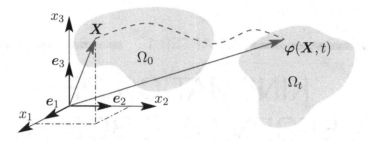

Figure 1.1: Motion from the reference configuration Ω_0 to the current configuration Ω_t.

esis of continuum mechanics: Matter is not discrete; it is continuously distributed in one-to-one correspondence with points in some subset of $\mathbb{R}^3$. Bodies are thus "continuous media": The components of X with respect to some basis are real numbers. Symbolically, we could write

$$\mathcal{B} = \{X\} \sim \{X = X_i e_i \in \overline{\Omega_0}\}$$

for some orthonormal basis $\{e_1, e_2, e_3\}$ and origin 0 chosen in three-dimensional Euclidean space and, thus, identified with $\mathbb{R}^3$. Hereafter, repeated indices are summed throughout their ranges; i.e. the "summation convention" is employed.

Kinematics is the study of the motion of bodies, without regard to the causes of the motion. It is purely a study of geometry and is an exact science within the hypothesis of a *continuum* (a continuous media).

1.1   Motion

We pick a point 0 in $\mathbb{R}^3$ as the origin of a fixed coordinate system $(x_1, x_2, x_3) = x$ defined by orthonormal vectors e_i, $i = 1, 2, 3$. The system (x_1, x_2, x_3) is called the *spatial* coordinate system. When the physical body $\mathcal{B}$ occupies its reference configuration Ω_0 at, say, time $t = 0$, the material point X occupies a position (place) corresponding to the vector $X = X_i e_i$. The spatial coordinates (X_1, X_2, X_3) of X

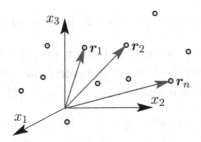

Figure 1.2: A discrete set of material particles.

are *labels* that identify the material point. The coordinate labels X_i are sometimes called *material* coordinates (see Fig. 1.1).

Remark Notice that if there were a *countable* set of discrete material points, such as one might use in models of molecular or atomistic dynamics, the particles (discrete masses) could be labeled using natural numbers $n \in \mathbb{N}$, as indicated in Fig. 1.2. But the particles (material points) in a continuum are not countable, so the use of a label of three real numbers for each particle corresponding to the coordinates of their position (at $t = 0$) in the reference configuration seems to be a very natural way to identify such particles. □

The body moves through $\mathcal{E}$ over a period of time and occupies a configuration $\Omega_t \subset \mathbb{R}^3$ at time t. Thus, material points X in $\overline{\Omega_0}$ (the closure of Ω_0) are mapped into positions x in $\overline{\Omega_t}$ by a smooth vector-valued mapping (see Fig. 1.1)

$$\boxed{x = \varphi(X, t).} \tag{1.1}$$

Thus, $\varphi(X, t)$ is the spatial position of the material point X at time t. The one-parameter family $\{\varphi(X, t)\}$ of positions is called the *trajectory* of X. We demand that φ be differentiable, injective, and orientation preserving. Then φ is called the *motion* of the body:

1. Ω_t is called the current configuration of the body.

2. φ is injective (except possibly at the boundary $\partial\Omega_0$ of Ω_0).

3. φ is *orientation preserving* (which means that the physical material cannot penetrate itself or reverse the orientation of material coordinates, which means that $\det \nabla\varphi(\boldsymbol{X}, t) > 0$).

Hereafter we will not explicitly show the dependence of φ and other quantities on time t unless needed; this time dependency is taken up later.

The vector field

$$\boldsymbol{u} = \varphi(\boldsymbol{X}) - \boldsymbol{X} \tag{1.2}$$

is the *displacement* of point $\boldsymbol{X}$. Note that

$$d\boldsymbol{x} = \nabla\varphi(\boldsymbol{X})\, d\boldsymbol{X} \qquad \left(\text{i.e., } dx_i = \frac{\partial\varphi_i}{\partial X_j}\, dX_j\right).$$

The tensor

$$\boldsymbol{F}(\boldsymbol{X}) = \nabla\varphi(\boldsymbol{X}) \tag{1.3}$$

is called the *deformation gradient*. Clearly,

$$\boldsymbol{F}(\boldsymbol{X}) = \boldsymbol{I} + \nabla\boldsymbol{u}(\boldsymbol{X}), \tag{1.4}$$

where $\boldsymbol{I}$ is the identity tensor and $\nabla\boldsymbol{u}$ is the *displacement gradient*.

Some Definitions

- A *deformation* is homogeneous if $\boldsymbol{F} = \boldsymbol{C} = \text{constant}$.

- A motion is *rigid* if it is the sum of a translation $\boldsymbol{a}$ and a rotation $\boldsymbol{Q}$:
$$\varphi(\boldsymbol{X}) = \boldsymbol{a} + \boldsymbol{Q}\boldsymbol{X},$$
where $\boldsymbol{a} \in \mathbb{R}^3$, $\boldsymbol{Q} \in \mathbb{O}^3_+$, with $\mathbb{O}^3_+$ the set of orthogonal matrices of order 3 with determinant equal to $+1$.

- As noted earlier, the fact that the motion is orientation preserving means that

$$\det \nabla \varphi(\boldsymbol{X}) > 0 \qquad \forall \boldsymbol{X} \in \overline{\Omega_0}.$$

- Recall that

$$\mathrm{Cof}\,\mathbf{F} = \text{cofactor matrix (tensor) of }\mathbf{F} = \det \mathbf{F}\,\mathbf{F}^{-T}.$$

For any matrix $\mathbf{A} = [A_{ij}]$ of order n and for each row i and column j, let $\mathbf{A}'_{ij}$ be the matrix of order $n-1$ obtained by deleting the ith row and jth column of $\mathbf{A}$. Let $d_{ij} = (-1)^{i+j} \det \mathbf{A}'_{ij}$. Then the matrix

$$\mathrm{Cof}\,\mathbf{A} = [d_{ij}]$$

is the cofactor matrix of $\mathbf{A}$ and d_{ij} is the (i,j) cofactor of $\mathbf{A}$. Note that

$$\boxed{\mathbf{A}(\mathrm{Cof}\,\mathbf{A})^T = (\mathrm{Cof}\,\mathbf{A})^T\mathbf{A} = (\det \mathbf{A})\mathbf{I}.} \qquad (1.5)$$

1.2 Strain and Deformation Tensors

A differential material line segment in the reference configuration is

$$dS_0^2 = d\boldsymbol{X}^T d\boldsymbol{X} = dX_1^2 + dX_2^2 + dX_3^2,$$

while the same material line in the current configuration is

$$dS^2 = d\boldsymbol{x}^T d\boldsymbol{x} = d\boldsymbol{X}^T \mathbf{F}^T \mathbf{F} d\boldsymbol{X}.$$

The tensor

$$\boxed{\mathbf{C} = \mathbf{F}^T\mathbf{F} = \text{the } \textit{right Cauchy--Green deformation tensor}}$$

is thus a measure of the change in dS_0^2 due to (gradients of) the motion

$$dS^2 - dS_0^2 = d\boldsymbol{X}^T\mathbf{C}d\boldsymbol{X} - d\boldsymbol{X}^T d\boldsymbol{X}.$$

C is symmetric, positive definite.   Another deformation measure is simply

$$dS^2 - dS_0^2 = d\boldsymbol{X}^T (2\mathbf{E})\, d\boldsymbol{X},$$

where

$$\mathbf{E} = \frac{1}{2}(\mathbf{C} - \mathbf{I}) = \text{the } \textit{Green–St. Venant strain tensor.} \tag{1.6}$$

Since $\mathbf{F} = \mathbf{I} + \nabla\boldsymbol{u}$ and $\mathbf{C} = \mathbf{F}^T\mathbf{F}$, we have

$$\mathbf{E} = \frac{1}{2}(\nabla\boldsymbol{u} + \nabla\boldsymbol{u}^T + \nabla\boldsymbol{u}^T\nabla\boldsymbol{u}). \tag{1.7}$$

The tensor

$$\mathbf{B} = \mathbf{F}\mathbf{F}^T = \text{the } \textit{left Cauchy–Green deformation tensor}$$

is also symmetric and positive definite, and we can likewise define

$$dS^2 - dS_0^2 = d\boldsymbol{X}^T\, \mathbf{F}^T(2\hat{\mathbf{A}})\mathbf{F}\, d\boldsymbol{X} = d\boldsymbol{x}^T(2\hat{\mathbf{A}})\, d\boldsymbol{x},$$

where

$$\hat{\mathbf{A}} = \frac{1}{2}(\mathbf{I} - \mathbf{B}^{-1}) = \text{the } \textit{Almansi–Hamel strain tensor} \tag{1.8}$$

or

$$\hat{\mathbf{A}} = \frac{1}{2}(\text{grad}\,\boldsymbol{u} + \text{grad}\,\boldsymbol{u}^T - \text{grad}\,\boldsymbol{u}\ \text{grad}\,\boldsymbol{u}^T), \tag{1.9}$$

where grad $\boldsymbol{u}$ is the spatial gradient grad $\boldsymbol{u} = \partial\boldsymbol{u}/\partial\boldsymbol{x}$ (i.e., $(\text{grad}\,\boldsymbol{u})_{ij} = \partial u_i(\boldsymbol{X}(\boldsymbol{x}), t)/\partial x_j$); see also Sec. 1.3.

Interpretation of E   Take $dS_0 = dX_1$ (i.e., $d\boldsymbol{X} = (dX_1, 0, 0)^T$). Then

$$dS^2 - dS_0^2 = dS^2 - dX_1^2 = 2E_{11}\, dX_1^2,$$

so

$$E_{11} = \frac{1}{2}\left(\left(\frac{dS}{dX_1}\right)^2 - 1\right) = \left\{\begin{array}{l} \text{a measure of the stretch of a} \\ \text{material line originally oriented} \\ \text{in the } X_1 \text{ direction in } \Omega_0. \end{array}\right.$$

We call e_1 the extension in the X_1 direction at $\boldsymbol{X}$ (which is a dimensionless measure of change in length per unit length)

$$e_1 \stackrel{\text{def}}{=} \frac{dS - dX_1}{dX_1} = \sqrt{1 + 2E_{11}} - 1,$$

or

$$2E_{11} = (1 + e_1)^2 - 1.$$

Similar definitions apply to E_{22} and E_{33}.

Now take $d\boldsymbol{X} = (dX_1, dX_2, 0)^T$ and

$$\cos\theta = \frac{d\boldsymbol{x}_1 \cdot d\boldsymbol{x}_2}{\|d\boldsymbol{x}_1\|\, \|d\boldsymbol{x}_2\|} = \frac{C_{12}}{\sqrt{1 + 2E_{11}}\sqrt{1 + 2E_{22}}} \qquad \text{(Exercise).}$$

The *shear* (or *shear strain*) in the X_1–X_2 plane is defined by the angle change (see Fig. 1.3),

$$\gamma_{12} \stackrel{\text{def}}{=} \frac{\pi}{2} - \theta.$$

Therefore

$$\boxed{\sin\gamma_{12} = \frac{2E_{12}}{\sqrt{1 + 2E_{11}}\sqrt{1 + 2E_{22}}}.} \qquad (1.10)$$

Thus, E_{12} (and, analogously, E_{13} and E_{23}) is a measure of the shear in the X_1–X_2 (or X_1–X_3 and X_2–X_3) plane.

Small strains   The tensor

$$\boldsymbol{e} = \frac{1}{2}(\nabla\boldsymbol{u} + \nabla\boldsymbol{u}^T), \qquad (1.11)$$

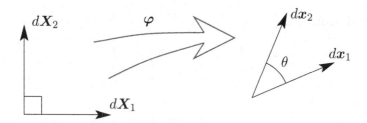

Figure 1.3: Change of angle through the motion φ.

is called the *infinitesimal* or *small* or *engineering strain tensor*. Clearly

$$\mathbf{E} = e + \frac{1}{2}\nabla u^T \nabla u. \tag{1.12}$$

Note that if $\mathbf{E}$ is "small" (i.e., $|E_{ij}| \ll 1$), then we obtain

$$\begin{aligned}
e_1 &= (1 + 2E_{11})^{1/2} - 1 \\
&= 1 + E_{11} - 1 + 0(E_{11}^2) \\
&\approx E_{11} = e_{11},
\end{aligned}$$

that is,

$$e_{11} = e_1 = \frac{dS - dX_1}{dX_1}, \quad \text{etc.,}$$

and

$$2e_{12} = \sin \gamma_{12} \approx \gamma_{12}, \quad \text{etc.}$$

Thus, small strains can be given the classical textbook interpretation: e_n is the change in length per unit length and e_{12} is the change in the right angle between material lines in the X_1 and X_2 directions. In the case of small strains, the Green–St. Venant strain tensor and the Almansi–Hamel strain tensor are indistinguishable.

1.3   Rates of Motion

If $\varphi(\mathbf{X}, t)$ is the motion (of $\mathbf{X}$ at time t), i.e.,

$$\boldsymbol{x} = \varphi(\mathbf{X}, t),$$

then

$$\dot{x} = \dot{x}(X, t) \stackrel{\text{def}}{=} \frac{\partial \varphi(X, t)}{\partial t} \tag{1.13}$$

is the velocity and

$$\ddot{x} = \ddot{x}(X, t) \stackrel{\text{def}}{=} \frac{\partial^2 \varphi(X, t)}{\partial t^2} \tag{1.14}$$

is the acceleration. Since φ is (in general) bijective, we can also describe the velocity as a function of the place x in $\mathbb{R}^3$ and time t:

$$v = v(x, t) = \dot{x}(\varphi^{-1}(x, t), t).$$

This is called the *spatial description* of the velocity.

This leads to two different ways to interpret the rates of motion of continua:

- The *material* description (functions are defined on *material points* X in the body $\mathcal{B}$ in correspondence to points in $\mathbb{R}^3$);

- The *spatial* description (functions are defined on (spatial) *places* x in $\mathbb{R}^3$).

When the equations of continuum mechanics are written in terms of the material description, the collective equations are commonly referred to as the *Lagrangian* form (formulation) of the equations (see Fig. 1.4). When the spatial description is used, the term *Eulerian* form (formulation) is used (see Fig. 1.5).

There are differences in the way rates of change appear in the Lagrangian and Eulerian formulations.

- In the Lagrangian case: Given a field $\psi_m = \psi_m(X, t)$ (the subscript m reminding us that we presume ψ is a function of the material coordinates),

$$\frac{d\psi_m(X, t)}{dt} = \frac{\partial \psi_m(X, t)}{\partial t} + \frac{\partial \psi_m(X, t)}{\partial X} \cdot \frac{\partial X}{\partial t},$$

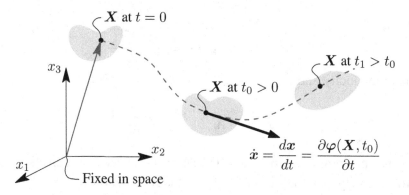

Figure 1.4: Lagrangian (material) description of velocity. The velocity of a material point is the time rate of change of the position of the point as it moves along its path (its trajectory) in $\mathbb{R}^3$.

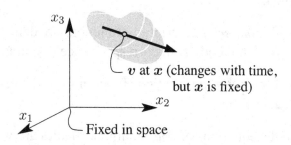

Figure 1.5: Eulerian (spatial) description of velocity. The velocity at a fixed place x in $\mathbb{R}^3$ is the speed and direction (at time t) of particles flowing through the place x.

but $\partial \boldsymbol{X}/\partial t = 0$ because $\boldsymbol{X}$ is simply a label of a material point. Thus,

$$\frac{d\boldsymbol{\psi}_m(\boldsymbol{X},t)}{dt} = \frac{\partial \boldsymbol{\psi}_m(\boldsymbol{X},t)}{\partial t}. \tag{1.15}$$

- In the Eulerian case: Given a field $\psi = \psi(x, t)$,

$$\frac{d\psi(x, t)}{dt} = \frac{\partial \psi(x, t)}{\partial t}\bigg|_{x \text{ fixed}} + \frac{\partial \psi(x, t)}{\partial x} \cdot \frac{\partial x}{\partial t},$$

but $\dfrac{\partial x}{\partial t} = v(x, t)$ is the velocity at position x and time t. Thus,

$$\frac{d\psi(x, t)}{dt} = \frac{\partial \psi(x, t)}{\partial t} + v(x, t) \cdot \frac{\partial \psi(x, t)}{\partial x}. \tag{1.16}$$

Notation   We distinguish between the gradient and divergence of fields in the Lagrangian and Eulerian formulation as follows:

Lagrangian: $\dfrac{\partial}{\partial X} = \nabla = \text{Grad},$	$\dfrac{\partial}{\partial X} \cdot \varphi = \nabla \cdot \varphi = \text{Div}\,\varphi;$
Eulerian: $\dfrac{\partial}{\partial x} = \text{grad},$	$\dfrac{\partial}{\partial x} \cdot v = \text{div}\,v.$

In classical literature, some authors write

$$\frac{D\psi}{Dt} = \frac{\partial \psi}{\partial t} + v \cdot \text{grad}\,\psi \tag{1.17}$$

as the "material time derivative" of a scalar field ψ, giving the rate of change of ψ at a fixed described place x at time t. Thus, in the Eulerian formulation, the acceleration is

$$a = \frac{Dv}{Dt} = \frac{\partial v}{\partial t} + (v \cdot \text{grad})\,v,$$

v being the velocity.

1.4   Rates of Deformation

The spatial (Eulerian) field

$$\mathbf{L} = \mathbf{L}(x, t) \stackrel{\text{def}}{=} \frac{\partial}{\partial x} v(x, t) = \text{grad}\,v(x, t) \tag{1.18}$$

is the *velocity gradient*. The time rate of change of the deformation gradient **F** is

$$\dot{\mathbf{F}} \equiv \frac{\partial}{\partial t} \nabla \varphi(\mathbf{X}, t) = \nabla \frac{\partial \varphi}{\partial t}(\mathbf{X}, t)$$

$$= \frac{\partial}{\partial \mathbf{X}} v(\mathbf{x}, t) = \frac{\partial v}{\partial \mathbf{x}} \frac{\partial \mathbf{x}}{\partial \mathbf{X}} = \operatorname{grad} v \mathbf{F},$$

or

$$\boxed{\dot{\mathbf{F}} = \operatorname{grad} v \mathbf{F} = \mathbf{L}_m \mathbf{F},} \tag{1.19}$$

where $\mathbf{L}_m = \mathbf{L}$ is written in material coordinates, so

$$\mathbf{L}_m = \dot{\mathbf{F}} \mathbf{F}^{-1}. \tag{1.20}$$

It is standard practice to write **L** in terms of its symmetric and skew-symmetric parts:

$$\boxed{\mathbf{L} = \mathbf{D} + \mathbf{W}.} \tag{1.21}$$

Here

$$\begin{aligned}
\mathbf{D} &= \frac{1}{2}(\mathbf{L} + \mathbf{L}^T) &= \text{ the *deformation rate tensor*,} \\
\mathbf{W} &= \frac{1}{2}(\mathbf{L} - \mathbf{L}^T) &= \text{ the *spin tensor*.}
\end{aligned} \tag{1.22}$$

We can easily show that if v is the velocity field,

$$\mathbf{W} v = \frac{1}{2}\, \boldsymbol{\omega} \times v, \tag{1.23}$$

where $\boldsymbol{\omega}$ is the vorticity

$$\boldsymbol{\omega} = \operatorname{curl} v. \tag{1.24}$$

Recall (cf. Exercise 2.6) that

$$D(\det \mathbf{A}) : \mathbf{V} = (\det \mathbf{A}) \mathbf{V}^T : \mathbf{A}^{-1},$$

for any invertible tensor **A** and arbitrary $\mathbf{V} \subset L(V, V)$. Also, if $f(g(t)) = f \circ g(t)$ denotes the composition of functions f and g, the chain rule of differentiation leads to

$$\frac{df(g(t))}{dt} = df(g(t)) \cdot \frac{dg(t)}{dt} = Df(g(t)) : \dot{g}(t).$$

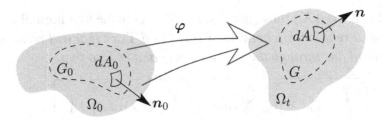

Figure 1.6: Mapping from reference configuration into current configuration.

Combining these expressions, we have

$$\overline{\det \mathbf{F}} = \frac{\partial \det \mathbf{F}}{\partial t} = D(\det \mathbf{F}) : \dot{\mathbf{F}} = \det \mathbf{F} \, \dot{\mathbf{F}}^T : \mathbf{F}^{-1}$$
$$= \det \mathbf{F} \, \operatorname{tr} \mathbf{L}_m = \det \mathbf{F} \, \operatorname{div} v$$

(since $\dot{\mathbf{F}}^T : \mathbf{F}^{-1} = \operatorname{tr} \dot{\mathbf{F}} \mathbf{F}^{-1} = \operatorname{tr} \mathbf{L}_m$, where $\operatorname{tr} \mathbf{L} = \operatorname{tr} \operatorname{grad} v = \operatorname{div} v$).
Summing up:

$$\boxed{\overline{\det \mathbf{F}} = \det \mathbf{F} \, \operatorname{div} v.} \tag{1.25}$$

There is a more constructive way of deriving (1.25) using the definitions of determinant and cofactors of $\mathbf{F}$; see Exercise 4 in Set I.2.

1.5   The Piola Transformation

The situation is this: A subdomain $G_0 \subset \Omega_0$ of the reference configuration of a body, with boundary ∂G_0 and unit exterior vector n_0 normal to the surface-area element dA_0, is mapped by the motion φ into a subdomain $G = \varphi(G_0) \subset \Omega_t$ of the current configuration with boundary ∂G with unit exterior vector n normal to the "deformed" surface area dA (see Fig. 1.6).

Let $\mathbf{T} = \mathbf{T}(x) = \mathbf{T}(\varphi(X))$ denote a tensor field defined on G and $\mathbf{T}(x)\,n(x)$ the flux of $\mathbf{T}$ across ∂G, $n(x)$ being a unit normal to ∂G. Here Ω_t is fixed so t is held constant and not displayed. Corresponding to $\mathbf{T}$, a tensor field $\mathbf{T}_0 = \mathbf{T}_0(X)$ is defined on G_0 that associates the flux $\mathbf{T}_0(\boldsymbol{X})\,\boldsymbol{n}_0(\boldsymbol{X})$ through ∂G_0, $\boldsymbol{n}_0(\boldsymbol{X})$ being the unit normal to ∂G_0. *We seek a relationship between $\mathbf{T}_0(\boldsymbol{X})$ and $\mathbf{T}(\boldsymbol{x})$ that will result in the same total flux through the surfaces ∂G_0 and ∂G, so that*

$$\int_{\partial G_0} \mathbf{T}_0(\boldsymbol{X})\boldsymbol{n}_0(\boldsymbol{X})\,dA_0 = \int_{\partial G} \mathbf{T}(\boldsymbol{x})\boldsymbol{n}(\boldsymbol{x})\,dA, \qquad (1.26)$$

with $\boldsymbol{x} = \boldsymbol{\varphi}(\boldsymbol{X})$. This relationship between $\mathbf{T}_0$ and $\mathbf{T}$ is called the *Piola transformation*.

Proposition 1.1 (Piola Transformation) *The above correspondence holds if*

$$\mathbf{T}_0(\boldsymbol{X}) = \det \mathbf{F}(\boldsymbol{X})\,\mathbf{T}(\boldsymbol{x})\,\mathbf{F}(\boldsymbol{X})^{-T} = \mathbf{T}(\boldsymbol{x})\,\operatorname{Cof}\mathbf{F}(\boldsymbol{X}). \quad (1.27)$$

Proof (This development follows that of Ciarlet [2]). We will use the Green's formulas (divergence theorems)

$$\int_{G_0} \operatorname{Div}\mathbf{T}_0\,dX = \int_{\partial G_0} \mathbf{T}_0\boldsymbol{n}_0\,dA_0$$

and

$$\int_{G} \operatorname{div}\mathbf{T}\,dx = \int_{\partial G} \mathbf{T}\boldsymbol{n}\,dA,$$

where

$$\operatorname{Div}\mathbf{T}_0 = \nabla \cdot \mathbf{T}_0 = \frac{\partial (\mathbf{T}_0)_{ij}}{\partial X_j}\boldsymbol{e}_i,$$

$$\operatorname{div}\mathbf{T} = \frac{\partial}{\partial x_j}T_{ij}\boldsymbol{e}_i,$$

$$dx = dx_1 dx_2 dx_3 = \det \mathbf{F}\,dX = \det \mathbf{F}\,dX_1 dX_2 dX_3.$$

We will also need to use the fact that

$$\frac{\partial}{\partial X_j}(\operatorname{Cof}\nabla\boldsymbol{\varphi})_{ij} = 0.$$

To show this, we first verify by direct calculation that

$$(\text{Cof } \mathbf{F})_{ij} = (\text{Cof } \nabla \boldsymbol{\varphi})_{ij} = \left(\frac{\partial}{\partial X_{j+i}} \varphi_{i+1} \right) \left(\frac{\partial}{\partial X_{j+2}} \varphi_{i+2} \right)$$
$$- \left(\frac{\partial}{\partial X_{j+2}} \varphi_{i+1} \right) \left(\frac{\partial}{\partial X_{j+1}} \varphi_{i+2} \right),$$

where no summation is used. Then a direct computation shows that

$$\frac{\partial}{\partial X_j} (\text{Cof } \mathbf{F})_{ij} = 0.$$

Next, set

$$\mathbf{T}_0(\boldsymbol{X}) = \mathbf{T}(\boldsymbol{x}) \, \text{Cof } \mathbf{F}(\boldsymbol{X}).$$

Noting that

$$\frac{1}{\det \mathbf{F}} (\text{Cof } \mathbf{F})^T = \mathbf{F}^{-1}$$

and

$$\frac{\partial x_i}{\partial X_m} \cdot \frac{\partial X_m}{\partial x_j} = \frac{\partial x_i}{\partial x_j} = \delta_{ij},$$

we see that

$$(\text{Cof } \mathbf{F})_{ij} = \det \mathbf{F} \, (\mathbf{F}^{-1})_{ji} = \det \mathbf{F} \, \frac{\partial X_j}{\partial x_i}.$$

Thus,

$$\text{Div } \mathbf{T}_0(\boldsymbol{X}) = e_k \frac{\partial}{\partial X_k} (\mathbf{T}_0)_{ij} e_i \otimes e_j = \frac{\partial (\mathbf{T}_0)_{ij}}{\partial X_j} e_i$$
$$= \frac{\partial}{\partial X_j} ((\mathbf{T}(\boldsymbol{x}) \, \text{Cof } \mathbf{F}(\boldsymbol{X}))_{ij} e_i$$
$$= \frac{\partial T_{im}(\boldsymbol{x})}{\partial x_r} \cdot \frac{\partial x_r}{\partial X_j} \cdot \text{Cof } \mathbf{F}(\boldsymbol{X})_{mj} e_i$$
$$+ \mathbf{T}_{im} \frac{\partial}{\partial X_j} \overset{0}{\cancel{\text{Cof } \mathbf{F}(\boldsymbol{X})_{mj}}} e_i$$
$$= \frac{\partial T_{im}}{\partial x_r} \cdot \frac{\partial x_r}{\partial X_j} \det \mathbf{F} \, \frac{\partial X_j}{\partial x_m} e_i$$
$$= \frac{\partial T_{ir}}{\partial x_r} e_i \det \mathbf{F}$$
$$= \text{div } \mathbf{T} \, \det \mathbf{F},$$

that is,

$$\boxed{\text{Div } \mathbf{T}_0 = \det \mathbf{F} \ \text{div } \mathbf{T}.}$$ (1.28)

Thus

$$\int_{G_0} \text{Div } \mathbf{T}_0 \, dX = \int_{G_0} \det \mathbf{F} \ \text{div } \mathbf{T} \, dX = \int_{\partial G_0} \mathbf{T}_0 \boldsymbol{n}_0 \, dA_0,$$

$$\int_{\partial G_0} \mathbf{T}_0 \boldsymbol{n}_0 \, dA_0 = \int_{G_0} \text{div } \mathbf{T} \det \mathbf{F} \, dX = \int_{G} \text{div } \mathbf{T} \, dx = \int_{\partial G} \mathbf{T}\boldsymbol{n} \, dA,$$

as asserted.

$\square$

Corollaries and Observations   The Piola transformation provides a means for characterizing the flux of a field through a material surface in the current configuration in terms of the representation of the surface in the reference configuration. It also provides fundamental relationships between differential surface areas and their orientations in the reference and current configurations. We list a few of these as corollaries and observations.

- Since G_0 is arbitrary (symbolically), we obtain

$$\boxed{\mathbf{T}_0 \boldsymbol{n}_0 \, dA_0 = \mathbf{T}\boldsymbol{n} \ dA.}$$ (1.29)

- Set $\mathbf{T} = \mathbf{I} = $ identity. Then

$$\boxed{\det \mathbf{F} \ \mathbf{F}^{-T} \boldsymbol{n}_0 \, dA_0 = \boldsymbol{n} \ dA.}$$ (1.30)

- Since $\boldsymbol{n} = \dfrac{dA_0}{dA} \cdot (\det \mathbf{F}) \mathbf{F}^{-T} \boldsymbol{n}_0$ and $\|\boldsymbol{n}\| = 1$, we have

$$\boxed{dA = \det \mathbf{F} \ \|\mathbf{F}^{-T} \boldsymbol{n}_0\| dA_0}$$ (Nanson's Formula), (1.31)

where $\| \cdot \|$ denotes the Eulerian norm. Thus

$$\boxed{\boldsymbol{n} = \frac{\text{Cof } \mathbf{F}\boldsymbol{n}_0}{\| \text{Cof } \mathbf{F}\boldsymbol{n}_0\|}.}$$ (1.32)

1.6 The Polar Decomposition Theorem

Theorem 1.A (Polar Decomposition) *A real invertible matrix* **F** *can be factored in a unique way as*

$$F = RU = VR, \tag{1.33}$$

where **R** *is an orthogonal matrix and* **U** *and* **V** *are symmetric positive definite matrices.*

Proof (We will use as a fact the following lemma: For every symmetric positive definite matrix **A**, there exists a unique symmetric positive definite matrix **B** such that $B^2 = A$.) Let us first show the existence of the matrices **U** and **V**. Define **U** by

$$U^2 = F^T F = C$$

(which is possible by virtue of the lemma stated above). Then let

$$R = FU^{-1}.$$

Then

$$R^T R = U^{-1} F^T F U^{-1} = U^{-1} U U U^{-1} = I.$$

Thus **R** is a rotation. We have thus shown that there exists a **U** such that $F = RU$.

Next, define

$$V = RUR^T.$$

Then

$$VR = RUR^T R = RU = F,$$

as asserted.

To show that **U** and **V** are unique, let $F = RU$, **R** being the rotation matrix. Then $F^T F = UR^T RU = U^2$, which means **U** is unique by the lemma stated. Since $R = FU^{-1}$, **R** is also uniquely defined. Finally, if $F = VR$, then $FF^T = B = V^2$, so by the same lemma, **V** is unique. □

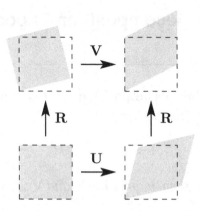

Figure 1.7: The Polar Decomposition Theorem: $\mathbf{F} = \mathbf{RU} = \mathbf{VR}$.

Summing up, if $\mathbf{C} = \mathbf{F}^T\mathbf{F}$ and $\mathbf{B} = \mathbf{FF}^T$ are the right and left Cauchy–Green deformation tensors and

$$\mathbf{F} = \mathbf{RU} \sim \mathbf{RU} = \mathbf{VR},$$

then

$$
\begin{aligned}
\mathbf{C} &= \mathbf{U}^T\mathbf{R}^T\mathbf{RU} = \mathbf{U}^T\mathbf{U} = \mathbf{U}^2, \\
\mathbf{B} &= \mathbf{VRR}^T\mathbf{V} = \mathbf{VV}^T = \mathbf{V}^2,
\end{aligned}
\tag{1.34}
$$

where $\mathbf{U}$ and $\mathbf{V}$ are the right and left stretch tensors, respectively.

Clearly, the Polar Decomposition Theorem establishes that the deformation gradient $\mathbf{F}$ can be obtained (or can be viewed) as the result of a distortion followed by a rotation or vice versa (see Fig. 1.7).

1.7 Principal Directions and Invariants of Deformation and Strain

For a given deformation tensor field $\mathbf{C}(\boldsymbol{X})$ and strain field $\mathbf{E}(\boldsymbol{X})$ (at point $\boldsymbol{X}$), recall that $d\boldsymbol{X}^T\mathbf{C}d\boldsymbol{X} = 2d\boldsymbol{X}^T\mathbf{E}d\boldsymbol{X} - d\boldsymbol{X}^Td\boldsymbol{X}$ is the square dS^2 of a material line segment in the current configuration. Suppose the material line in question is oriented in the direction of a unit vector m in the reference configuration so that $dX = m \, dS_0$. Then a measure of the stretch or compression of a unit material element originally oriented along a unit vector m is given by

$$\Delta(m) = dS^2/dS_0^2 = m^T C m,$$
$$m \cdot m = \text{“}m^T m\text{”} = 1.$$

(1.35)

One may ask: Of all possible directions m at X, which choice results in the largest (or smallest) value of $\Delta(m)$?

This is a constrained maximization/minimization problem: Find $m = m_{\max}$ (or $m_{\min}$) that makes $\Delta(m)$ as large (or small) as possible, subject to the constraint $m^T m = 1$. To resolve this problem, we use the method of Lagrange multipliers. Denote by $L(m, \lambda) = \Delta(m) - \lambda(m^T m - 1)$, λ being the Lagrange multiplier. The maxima (on minimize and maximize points) of L satisfy,

$$\frac{\partial L(m, \lambda)}{\partial m} = 0 = 2(C m - \lambda m).$$

Thus, unit vectors m that maximize or minimize $\Delta(m)$ are associated with multipliers λ and satisfy

$$C m = \lambda m, \qquad m^T m = 1.$$

(1.36)

That is, (m, λ) are eigenvector/eigenvalue pairs of the deformation tensor C, and m is normalized so that $m^T m = 1$ (or $\|m\| = 1$).

The following fundamental properties of the above eigenvalue problem can be listed.

1. There are three real eigenvalues and three eigenvectors of C (at X); we adopt the ordering $\lambda_1 \geq \lambda_2 \geq \lambda_3$.

2. For $\lambda_i \neq \lambda_j$, the corresponding eigenvectors are orthogonal (for pairs (m_i, λ_i) and (m_j, λ_j)), $m_i^T m_j = \delta_{ij}$, as can be seen as follows:

$$m_i^T(\lambda_j m_j) = m_i^T C m_j = m_j^T C m_i = m_j^T(\lambda_i m_i),$$

so

$$(\lambda_i - \lambda_j)\boldsymbol{m}_i^T \boldsymbol{m}_j = 0,$$

so

$$\text{if } \lambda_i \neq \lambda_j, \qquad \boldsymbol{m}_i^T \boldsymbol{m}_j = \delta_{ij}, \ 1 \leq i, j \leq 3$$

(if $\lambda_i = \lambda_j$, we can always construct $\boldsymbol{m}_j$ so that it is orthogonal to $\boldsymbol{m}_i$).

3. Equation (1.36) can be written as

$$(\mathbf{C} - \lambda\mathbf{I})\boldsymbol{m} = \mathbf{0}. \tag{1.37}$$

This equation can have nontrivial solutions only if the determinant of $\mathbf{C} - \lambda\mathbf{I}$ is zero. This is precisely the *characteristic polynomial* of $\mathbf{C}$:

$$\det(\mathbf{C} - \lambda\mathbf{I}) = -\lambda^3 + I(\mathbf{C})\lambda^2 - I\!I(\mathbf{C})\lambda + I\!I\!I(\mathbf{C}), \tag{1.38}$$

where $I, I\!I, I\!I\!I$ are the *principal invariants* of $\mathbf{C}$:

$$
\begin{aligned}
I(\mathbf{C}) &= \text{trace } \mathbf{C} \equiv \text{tr } \mathbf{C} = C_{ii} = C_{11} + C_{22} + C_{33}, \\
I\!I(\mathbf{C}) &= \frac{1}{2}(\text{tr } \mathbf{C})^2 - \frac{1}{2}\text{tr } \mathbf{C}^2 \qquad = \text{tr Cof } \mathbf{C}, \\
I\!I\!I(\mathbf{C}) &= \det \mathbf{C} = \frac{1}{6}\left((\text{tr } \mathbf{C})^3 - 3\,\text{tr }\mathbf{C}\,\text{tr }\mathbf{C}^2 + 2\,\text{tr }\mathbf{C}^3\right).
\end{aligned}
\tag{1.39}
$$

(An invariant of a real matrix $\mathbf{C}$ is any real-valued function $\mu(\mathbf{C})$ with the property $\mu(\mathbf{C}) = \mu(\mathbf{A}^{-1}\mathbf{C}\mathbf{A})$ for all invertible matrices $\mathbf{A}$.)

4. Because the eigenvectors are all positive, it is customary to write λ_i^2 for the eigenvectors instead of λ_i. Then $(\mathbf{C} - \lambda_i^2\mathbf{I})\boldsymbol{m}_i = \mathbf{0}$. Let $\mathbf{N}$ be the matrix with the mutually orthogonal eigenvectors as rows. Then

$$\mathbf{N}^T\mathbf{C}\mathbf{N} = \begin{bmatrix} \lambda_1^2 & 0 & 0 \\ 0 & \lambda_2^2 & 0 \\ 0 & 0 & \lambda_3^2 \end{bmatrix} = \text{diag}\{\lambda_i^2, i = 1, 2, 3\}. \tag{1.40}$$

The coordinate system defined by the mutually orthogonal triad of eigenvectors define the principal directions and values of C at $\boldsymbol{X}$. For this choice of a basis, we obtain

$$\boldsymbol{C} = \sum_{i=1}^{3} \lambda_i^2 \boldsymbol{m}_i \otimes \boldsymbol{m}_i. \tag{1.41}$$

If $\lambda_1^2 \geq \lambda_2^2 \geq \lambda_3^2$, λ_1^2 corresponds to the maximum, λ_3^2 to the minimum, and λ_2^2 to a "mini-max" principal value of $\boldsymbol{C}$ (or of $\Delta(\boldsymbol{n})$).

Notice that the stretch along, say, $\boldsymbol{m}_1$ is $(\boldsymbol{m}_1^T \boldsymbol{C} \boldsymbol{m}_1)^{1/2} = \lambda_1$, etc. Also,

$$\boldsymbol{C} = \boldsymbol{U}^2 = \begin{bmatrix} \lambda_1^2 & 0 & 0 \\ 0 & \lambda_2^2 & 0 \\ 0 & 0 & \lambda_3^2 \end{bmatrix}. \tag{1.42}$$

The principal invariants are thus

$$\begin{aligned} I(\boldsymbol{C}) &= \lambda_1^2 + \lambda_2^2 + \lambda_3^2, \\ I\!I(\boldsymbol{C}) &= \lambda_1^2 \lambda_2^2 + \lambda_1^2 \lambda_3^2 + \lambda_2^2 \lambda_3^2, \\ I\!I\!I(\boldsymbol{C}) &= \lambda_1^2 \lambda_2^2 \lambda_3^2. \end{aligned} \tag{1.43}$$

1.8 The Reynolds' Transport Theorem

We frequently encounter the need to evaluate the total time rate of change of a field, either densities or measures of concentrations per unit volume, defined over a volume $\omega \subset \Omega_t$. For instance, if $\Psi = \Psi(\boldsymbol{x}, t)$ is a spatial field, either scalar- or vector-valued, suppose we wish to compute $d(\int_\omega \Psi dx)/dt$. The following change of integration variables facilitates such a calculation. Let ω_0 be the region of Ω_0 occupied by the material, while in the reference configuration, that occupies ω in Ω_t. Then,

$$
\begin{aligned}
\frac{d}{dt} \int_\omega \Psi \, dx &= \frac{d}{dt} \int_{\omega_0} \Psi_m \det \mathbf{F} \, dX \\
&= \int_{\omega_0} \frac{d}{dt} \left(\Psi_m \det \mathbf{F} \right) dX \\
&= \int_{\omega_0} \left(\frac{\partial \Psi_m}{\partial t} + \boldsymbol{v} \cdot \operatorname{grad} \Psi_m \right) \det \mathbf{F} \, dX \\
&\quad + \int_{\omega_0} \Psi_m \overline{\det \mathbf{F}} \det \mathbf{F} \, dX \\
&= \int_\omega \frac{\partial \Psi}{\partial t} dx + \int_\omega \operatorname{div}(\Psi \boldsymbol{v}) \, dx.
\end{aligned}
$$

Thus,

$$
\boxed{\frac{d}{dt} \int_\omega \Psi \, dx = \int_\omega \frac{\partial \Psi}{\partial t} dx + \int_{\partial \omega} \Psi \boldsymbol{v} \cdot \boldsymbol{n} \, dA.} \tag{1.44}
$$

This last result is known as the *Reynolds' Transport Theorem.*

CHAPTER 2

MASS AND MOMENTUM

A common dictionary definition of mass is as follows:

Mass *The property of a body that is a measure of the amount of material it contains and causes it to have weight in a gravitational field.*

In continuum mechanics, the mass of a body is continuously distributed over its volume and is an integral of a density field $\varrho : \overline{\Omega_0} \to \mathbb{R}^+$ called the mass density. The total mass $\mathcal{M}(\mathcal{B})$ of a body is independent of the motion φ, but the mass density ϱ can, of course, change as the volume of the body changes while in motion. Symbolically,

$$\mathcal{M}(\mathcal{B}) = \int_{\Omega_t} \varrho \, dx, \tag{2.1}$$

where dx = volume element in the current configuration Ω_t of the body.

Given two motions φ and ψ (see Fig. 2.1), let ϱ_φ and ϱ_ψ denote the mass densities in the configurations $\varphi(\Omega_0)$ and $\psi(\Omega_0)$, respectively.

An Introduction to Mathematical Modeling: A Course in Mechanics, First Edition. By J. Tinsley Oden
© 2011 John Wiley & Sons, Inc. Published 2011 by John Wiley & Sons, Inc.

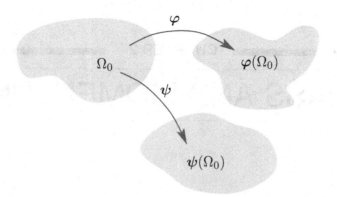

Figure 2.1: Two motions φ and ψ.

Since the total mass is independent of the motion,

$$
\mathcal{M}(\mathcal{B}) = \int_{\varphi(\Omega_0)} \varrho_\varphi \, dx = \int_{\psi(\Omega_0)} \varrho_\psi \, dx. \tag{2.2}
$$

This fact represents the *principle of conservation of mass*. The mass of a body $\mathcal{B}$ is thus an invariant property (measuring the amount of material in $\mathcal{B}$); the weight of $\mathcal{B}$ is defined as $g\mathcal{M}(\mathcal{B})$ where g is a constant gravity field. Thus, a body may weigh differently in different gravity fields (e.g., the earth's gravity as opposed to that on the moon), but its mass is the same.

2.1   Local Forms of the Principle of Conservation of Mass

Let $\varrho_0 = \varrho_0(\boldsymbol{X})$ be the mass density of a body in its reference configuration and let $\varrho = \varrho(\boldsymbol{x}, t)$ be the mass density in the current configuration Ω_t. Then

$$
\int_{\Omega_0} \varrho_0(\boldsymbol{X}) \, dX = \int_{\Omega_t} \varrho(\boldsymbol{x}) \, dx
$$

(where the dependence of ϱ on t has been suppressed). But $dx = \det \mathbf{F}(\mathbf{X}) \, dX$, so

$$\int_{\Omega_0} [\varrho_0(\mathbf{X}) - \varrho(\boldsymbol{\varphi}(\mathbf{X})) \det \mathbf{F}(\mathbf{X})] \, dX = 0,$$

and therefore

$$\varrho_0(\mathbf{X}) = \varrho(\mathbf{x}) \det \mathbf{F}(\mathbf{X}). \tag{2.3}$$

This is the *material description* (or the *Lagrangian formulation*) of the principle of conservation of mass. To obtain the *spatial description* (or *Eulerian formulation*), we observe that the invariance of total mass can be expressed as

$$\frac{d}{dt} \int_{\Omega_t} \varrho(\mathbf{x}, t) \, dx = 0.$$

Changing to the material coordinates gives

$$0 = \frac{d}{dt} \int_{\Omega_0} \varrho(\mathbf{x}, t) \det \mathbf{F}(\mathbf{X}, t) \, dX = \int_{\Omega_0} (\varrho \, \overline{\det \mathbf{F}} + \dot{\varrho} \det \mathbf{F}) \, dX,$$

where $(\dot{\cdot}) = d(\cdot)/dt$. Recalling that $\overline{\det \mathbf{F}} = \det \mathbf{F} \operatorname{div} \boldsymbol{v}$, we have

$$
\begin{aligned}
0 &= \int_{\Omega_0} \det \mathbf{F} \left(\varrho \operatorname{div} \boldsymbol{v} + \frac{\partial \varrho}{\partial t} + \boldsymbol{v} \cdot \operatorname{grad} \varrho \right) dX \\
&= \int_{\Omega_0} \left(\frac{\partial \varrho}{\partial t} + \operatorname{div}(\varrho \boldsymbol{v}) \right) \det \mathbf{F} \, dX \\
&= \int_{\Omega_t} \left(\frac{\partial \varrho}{\partial t} + \operatorname{div}(\varrho \boldsymbol{v}) \right) dx,
\end{aligned}
$$

from which we conclude

$$\frac{\partial \varrho}{\partial t} + \operatorname{div}(\varrho \boldsymbol{v}) = 0. \tag{2.4}$$

2.2 Momentum

The momentum of a material body is a property the body has by virtue of its mass and its velocity. Given a motion φ of a body $\mathcal{B}$ of mass density ϱ, the *linear momentum* $I(\mathcal{B}, t)$ of $\mathcal{B}$ at time t and the *angular momentum* $H(\mathcal{B}, t)$ of $\mathcal{B}$ at time t about the origin 0 of the spatial coordinate system are defined by

$$
\begin{aligned}
I(\mathcal{B}, t) &= \int_{\Omega_t} \varrho \boldsymbol{v} \, dx, \\
H(\mathcal{B}, t) &= \int_{\Omega_t} \boldsymbol{x} \times \varrho \boldsymbol{v} \, dx.
\end{aligned}
\tag{2.5}
$$

Again, $dx \ (= dx_1 dx_2 dx_3)$ is the volume element in Ω_t.

The rates of change of momenta (both I and H) are of fundamental importance. To calculate rates, first notice that for any smooth field $w = w(\boldsymbol{x}, t)$,

$$
\begin{aligned}
\frac{d}{dt} \int_{\Omega_t} w \varrho \, dx &= \frac{d}{dt} \int_{\Omega_0} w(\varphi(\boldsymbol{X}, t), t) \varrho(\boldsymbol{x}, t) \det \mathbf{F}(\boldsymbol{X}, t) \, dX \\
&= \int_{\Omega_0} \frac{dw}{dt} \varrho_0 \, dX = \int_{\Omega_t} \frac{dw}{dt} \varrho \, dx.
\end{aligned}
\tag{2.6}
$$

Thus,

$$
\begin{aligned}
\frac{dI(\mathcal{B}, t)}{dt} &= \int_{\Omega_t} \varrho \frac{d\boldsymbol{v}}{dt} \, dx, \\
\frac{dH(\mathcal{B}, t)}{dt} &= \int_{\Omega_t} \boldsymbol{x} \times \varrho \frac{d\boldsymbol{v}}{dt} \, dx.
\end{aligned}
\tag{2.7}
$$

CHAPTER 3

FORCE AND STRESS IN DEFORMABLE BODIES

The concept of force is used to characterize the interaction of the motion of a material body with its environment. More generally, as will be seen later, force is a characterization of interactions of the body with agents that cause a change in its momentum. In continuum mechanics, there are basically two types of forces: (1) *contact forces*, representing the contact of the boundary surfaces of the body with the exterior universe (i.e., its exterior environment) or the contact of internal parts of the body on surfaces that separate them, and (2) *body forces*, acting on material points of the body by its environment.

Body Forces Examples of body forces are the weight-per-unit volume exerted by the body by gravity or forces per unit volume exerted by an external magnetic field. Body forces are a type of *external force*, naturally characterized by a given vector-valued field f called the *body force density per unit volume*. The total body force is then

$$\int_{\Omega_t} f(x, t) \, dx.$$

An Introduction to Mathematical Modeling: A Course in Mechanics, First Edition. By J. Tinsley Oden
© 2011 John Wiley & Sons, Inc. Published 2011 by John Wiley & Sons, Inc.

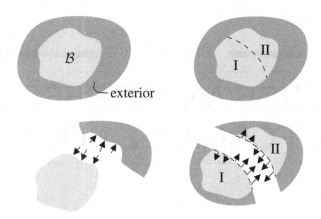

Figure 3.1: External contact forces (bottom left) and internal contact forces (bottom right).

Alternatively, we can measure the body force with a density b per unit mass: $f = \varrho b$. Then

$$\int_{\Omega_t} f \, dx = \int_{\Omega_t} \varrho b \, dx. \tag{3.1}$$

Contact Forces   Contact forces are also called *surface forces* because the contact of one body with another or with its surroundings must take place on a material surface. Contact forces fall into two categories (see Fig. 3.1):

- *External contact forces* representing the contact of the exterior boundary surface of the body with the environment outside the body, and

- *Internal contact forces* representing the contact of arbitrary parts of the body that touch one another on parts of internal surfaces they share on their common boundary.

The Concept of Stress   There is essentially no difference between the structure of external or internal contact forces; they differ only in what

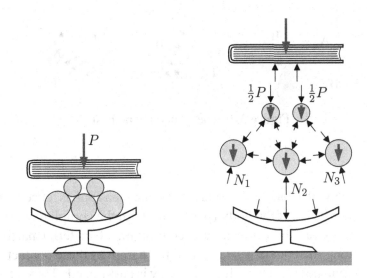

Figure 3.2: Illustrative example of the stress concept.

is *interpreted* as the boundary that separates a material body from its surroundings. Portion I of the partitioned body in Fig. 3.1 could just as well be defined as body $\mathcal{B}$ and portion II would then be part of its exterior environment.

Figure 3.2 is an illustration of the discrete version of the various forces: a collection of rigid spherical balls of weight W, each resting in a rigid bowl and pushed downward by balancing a book of weight P on the top balls. Explode the collection of balls into free bodies as shown. The five balls are the body $\mathcal{B}$. The exterior contact with the outside environment is represented by the force of magnitude P distributed into two equal parts of magnitude $\frac{1}{2}P$ and the contact forces N_i representing the fact that the balls press against the bowl and the bowl against the balls in an equal and opposite way. Then $\frac{1}{2}\mathbf{P}$ ($2\times$), $\mathbf{N}_1$, $\mathbf{N}_2$, and $\mathbf{N}_3$ are external contact forces. The weights $\mathbf{W}$ are the body forces. Internally, the balls touch one another on exterior surfaces of each ball. The action of a given ball on another is equal and opposite to the action of the other balls on the given ball. These contact forces are internal. They cancel out (balance) when the balls are reassembled into the whole body $\mathcal{B}$.

Figure 3.3: The Cauchy hypothesis.

In the case of a continuous body, the same idea applies, except that the contact of any part of the body (part I say) with the complement (part II) is continuous (as there are now a continuum of material particles in contact along the contact surface) and the nature of these contact forces depends upon how (we visualize) the body is partitioned. Thus, at a point x, if we separate $\mathcal{B}$ (conceptually) into bodies I and II with a surface AA defined with an orientation given by a unit vector n, the distribution of contact forces at a point x on the surface will be quite different than that produced by a different partitioning of the body defined by a different surface BB though the same point x but with orientation defined by a different unit vector m (see Fig. 3.3).

These various possibilities are captured by the so-called *Cauchy hypothesis*: There exists a vector-valued surface (contact) force density

$$\boldsymbol{\sigma}(n, x, t)$$

giving the force per unit area on an oriented surface Γ through x with unit normal n, at time t. The convention is that $\boldsymbol{\sigma}(n, x, t)$ defines the force per unit area on the "negative" side of the material (n is a unit exterior or outward normal) exerted by the material on the opposite side (thus, the direction of $\boldsymbol{\sigma}$ on body II is opposite to that on I because the exterior normals are in opposite directions; see Fig. 3.4).

Thus, if the vector field $\boldsymbol{\sigma}(n, x, t)$ were known, one could pick an arbitrary point x in the body (or, equivalently, in the current configuration Ω_t) at time t, and pass a surface through x with orientation given by the unit normal n. The vector $\boldsymbol{\sigma}(n, x, t)$ would then represent the contact force per unit area on this surface at point x at time t. The surface Γ through x partitions the body into two parts: The orientation of the

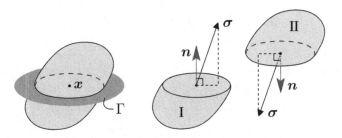

Figure 3.4: The stress vector $\boldsymbol{\sigma}$.

vector $\boldsymbol{\sigma}(\boldsymbol{n}, \boldsymbol{x}, t)$ on one part (at $\boldsymbol{x}$) is opposite to that on the other part. The vector field $\boldsymbol{\sigma}$ is called the *stress vector field* and $\boldsymbol{\sigma}(\boldsymbol{n}, \boldsymbol{x}, t)$ is the *stress vector* at $\boldsymbol{x}$ and t for orientation $\boldsymbol{n}$. The total force on surface Γ is

$$\int_\Gamma \boldsymbol{\sigma}(\boldsymbol{n}, \boldsymbol{x}, t)\, dA,$$

dA being the surface area element.

The total force acting on body $\mathcal{B}$ and the total moment about the origin $\mathbf{0}$, at time t when the body occupies the current configuration Ω_t, are, respectively,

$$
\begin{aligned}
\mathcal{F}(\mathcal{B}, t) &= \int_{\Omega_t} \boldsymbol{f}\, dx + \int_{\partial\Omega_t} \boldsymbol{\sigma}(\boldsymbol{n})\, dA, \\
\mathbf{M}(\mathcal{B}, t) &= \int_{\Omega_t} \boldsymbol{x} \times \boldsymbol{f}\, dx + \int_{\partial\Omega_t} \boldsymbol{x} \times \boldsymbol{\sigma}(\boldsymbol{n})\, dA,
\end{aligned}
\tag{3.2}
$$

where we have suppressed the dependence of $\boldsymbol{f}$ and $\boldsymbol{\sigma}$ on $\boldsymbol{x}$ and t.

CHAPTER 4

THE PRINCIPLES OF BALANCE OF LINEAR AND ANGULAR MOMENTUM

The momentum balance laws are the fundamental axioms of mechanics that connect motion and force:

The Principle of Balance of Linear Momentum *The time rate of change of linear momentum $I(\mathcal{B}, t)$ of a body $\mathcal{B}$ at time t equals (or is balanced by) the total force $\mathcal{F}(\mathcal{B}, t)$ acting on the body:*

$$\frac{dI(\mathcal{B}, t)}{dt} = \mathcal{F}(\mathcal{B}, t). \tag{4.1}$$

An Introduction to Mathematical Modeling: A Course in Mechanics, First Edition. By J. Tinsley Oden
© 2011 John Wiley & Sons, Inc. Published 2011 by John Wiley & Sons, Inc.

The Principle of Balance of Angular Momentum   *The time rate of change of angular momentum $H(\mathcal{B}, t)$ of a body $\mathcal{B}$ at time t equals (or is balanced by) the total moment $\mathbf{M}(\mathcal{B}, t)$ acting on the body:*

$$\frac{dH(\mathcal{B}, t)}{dt} = \mathbf{M}(\mathcal{B}, t). \tag{4.2}$$

Thus, for a continuous media,

$$\int_{\Omega_t} \varrho \frac{d\boldsymbol{v}}{dt}\, dx = \int_{\Omega_t} \mathbf{f}\, dx + \int_{\partial\Omega_t} \boldsymbol{\sigma}(\boldsymbol{n})\, dA,$$

$$\int_{\Omega_t} \boldsymbol{x} \times \varrho \frac{d\boldsymbol{v}}{dt}\, dx = \int_{\Omega_t} \boldsymbol{x} \times \mathbf{f}\, dx + \int_{\partial\Omega_t} \boldsymbol{x} \times \boldsymbol{\sigma}(\boldsymbol{n})\, dA. \tag{4.3}$$

4.1 Cauchy's Theorem: The Cauchy Stress Tensor

Theorem 4.A (Cauchy) *At each time t, let the body force density $\mathbf{f} : \Omega_t \to \mathbb{R}^3$ be a continuous function of $\boldsymbol{x}$ and let the stress vector field $\boldsymbol{\sigma} = \boldsymbol{\sigma}(\boldsymbol{n}, \boldsymbol{x}, t)$ be continuously differentiable with respect to $\boldsymbol{n}$ for each $\boldsymbol{x} \in \Omega_t$ and continuously differentiable with respect to $\boldsymbol{x}$ for each $\boldsymbol{n}$. Then the principles of balance of linear and angular momentum imply that there exists a continuously differential tensor field $\mathbf{T} : \overline{\Omega_t} \to \mathbb{M}^3$ (the set of square matrices of order three) such that*

$$\boldsymbol{\sigma}(\boldsymbol{n}, \boldsymbol{x}, t) = \mathbf{T}(\boldsymbol{x}, t)\boldsymbol{n}, \qquad \forall \boldsymbol{x} \in \overline{\Omega_t},\ \forall \boldsymbol{n} \tag{4.4}$$

and

$$\mathbf{T}(\boldsymbol{x}, t) = \mathbf{T}(\boldsymbol{x}, t)^T, \qquad \forall \boldsymbol{x} \in \overline{\Omega_t}. \tag{4.5}$$

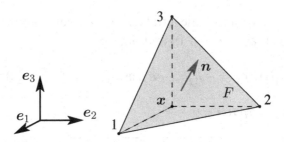

Figure 4.1: Tetrahedron τ for the proof of Cauchy's Theorem.

Proof The proof is classical: Pick $x \in \Omega_t$. Since Ω_t is open, we can construct a tetrahedron τ with "vertex" at x with three faces parallel to the x_i-coordinate planes and with an exterior face F with unit normal $n = n_i \, e_i, n_i > 0$ (see Fig. 4.1).

The faces of τ opposite to the vertices i, $i = 1, 2, 3$, are denoted F_i and area$(F_i) = n_i$area(F). According to the principle of balance of linear momentum, we have

$$\int_\tau \mathbf{f} \, dx + \int_{\partial\tau} \boldsymbol{\sigma}(n) \, dA - \int_\tau \varrho \frac{dv}{dt} \, dx = 0.$$

Let $\mathbf{f} = f_i e_i$, $\boldsymbol{\sigma}(n) = \sigma_i(n)e_i$, $v = v_i e_i$. Then, for each component of the surface integral term in the above equation, and using the mean-value theorem,

$$\int_{\partial\tau} \sigma_i(n, x) \, dA = \sum_{j=1}^3 \int_{F_j = Fn_j} \sigma_i(-e_j, x) \, dA + \int_F \sigma_i(n, x) \, dA$$

$$= \sigma_i(-e_j, x_j^*) \, n_j \, \text{area}(F) + \sigma_i(n, \widehat{x}) \, \text{area}(F),$$

for $x_j^* \in F_j$ and $\widehat{x} \in F$, so that

$$\int_{\partial\tau} \sigma_i(n, x) \, dA = \int_\tau \left(\varrho \frac{dv_i(x)}{dt} - f_i(x) \right) \, dx, \qquad i = 1, 2, 3.$$

Thus, since volume$(\tau) = C$ area$(F)^{3/2}$, C a constant and $i = 1, 2, 3$,

$$\left| n_j \sigma_i(-e_j, x_j^*) + \sigma_i(n, \widehat{x}) \right| \text{area}(F) \leq C \sup_{x \in \tau} \left| \varrho(x) \frac{dv_i(x)}{dt} - f_i(x) \right|$$

$$\times \, \text{area}(F)^{3/2}.$$

Keeping n fixed, we shrink the tetrahedron τ to the vertex x by collapsing the vertices to x (area$(F) \to 0$) and obtain

$$\sigma_i(n, x) = -n_j\sigma_i(-e_j, x), \qquad 1 \le i, j \le 3,$$

or, since $\sigma(n) = \sigma_i(n)e_i$,

$$\boxed{\sigma(n, x) = -n_j\sigma(-e_j, x).} \tag{4.6}$$

Now, for each vector $\sigma(e_j, x)$, define functions $T_{ij}(x)$ such that

$$T_{ij}(x) = -\sigma_i(-e_j, x).$$

Then $\sigma(e_j, x) = T_{ij}(x)e_i$, or

$$\boxed{\sigma_i(n, x) = T_{ij}(x)n_j, \quad \text{or} \quad \sigma(n, x) = \mathbf{T}(x)n,} \tag{4.7}$$

as asserted (by continuity, these hold for all $x \in \overline{\Omega_t}$). The tensor $\mathbf{T}$ is of course the Cauchy stress tensor. We will take up the proof that $\mathbf{T}$ is symmetric later (as an exercise), which follows from the principle of balance of angular momentum. □

The Cauchy stress tensor provides for a convenient bookkeeping scheme for representing components of the stress vector. Let $n = (1, 0, 0)$. Then $\sigma(n) = (T_{11}, T_{21}, T_{31}) = \sigma_1$ is the vector acting on a plane normal to the x_1-direction, with components T_{1i}, as indicated in Fig. 4.2. Similarly, the components T_{2i} and T_{3i} are components of the stress vector normal to the x_2 and x_3 planes, respectively.

4.2 The Equations of Motion (Linear Momentum)

According to the divergence theorem, we have

$$\int_{\partial\Omega_t} \mathbf{T}n \, dA = \int_{\Omega_t} \operatorname{div} \mathbf{T} \, dx.$$

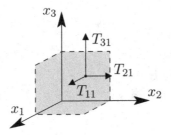

Figure 4.2: Illustration of the stress vector $\sigma_1 = (T_{11}, T_{21}, T_{31})$.

Thus,

$$\int_{\partial\Omega_t} \sigma(n)\, dA = \int_{\partial\Omega_t} \mathbf{T}n\, dA = \int_{\Omega_t} \operatorname{div} \mathbf{T}\, dx.$$

It follows that the principle of balance of linear momentum can be written,

$$\int_{\Omega_t} \varrho\frac{dv}{dt}dx = \int_{\Omega_t} (f + \operatorname{div} \mathbf{T})dx,$$

where $\mathbf{T}$ is the Cauchy stress tensor. Thus

$$\int_{\Omega_t} (\operatorname{div} \mathbf{T} + f - \varrho\frac{dv}{dt})dx = 0.$$

But this must also hold for any arbitrary subdomain $G \subset \Omega_t$. Thus,

$$\boxed{\operatorname{div} \mathbf{T} + f = \varrho\frac{dv}{dt},} \tag{4.8}$$

or

$$\operatorname{div} \mathbf{T}(x,t) + f(x,t) = \varrho(x,t)\frac{dv}{dt}(x,t).$$

Returning to the proof of Cauchy's Theorem, we apply the principle of balance of angular momentum to the tetrahedron and use the fact that $\operatorname{div} \mathbf{T} + f - \varrho dv/dt = 0$. This leads to the conclusion that

$$\boxed{\mathbf{T}^T = \mathbf{T}.}$$

Details are left as an exercise; see Set I.4.

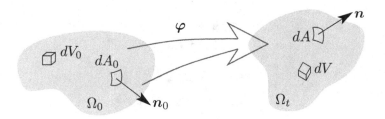

Figure 4.3: Mapping of volume and surface elements.

4.3 The Equations of Motion Referred to the Reference Configuration: The Piola–Kirchhoff Stress Tensors

In the current configuration, we have

$$\operatorname{div} \mathbf{T}(\boldsymbol{x}, t) + \boldsymbol{f}(\boldsymbol{x}, t) = \varrho(\boldsymbol{x}, t) \left(\frac{\partial \boldsymbol{v}(\boldsymbol{x}, t)}{\partial t} + \boldsymbol{v}(\boldsymbol{x}, t) \cdot \operatorname{grad} \boldsymbol{v}(\boldsymbol{x}, t) \right).$$
(4.9)

Here div and grad are defined with respect to $\boldsymbol{x}$; i.e. the dependent variables are regarded as functions of spatial position $\boldsymbol{x}$ (and t). We now refer the fields to the reference configuration (see Fig. 4.3):

$$\boldsymbol{f}_0(\boldsymbol{X}) = \boldsymbol{f}(\boldsymbol{x}) \det \mathbf{F}(\boldsymbol{X}) \qquad (\boldsymbol{x} = \boldsymbol{\varphi}(\boldsymbol{X})),$$

$$\varrho_0(\boldsymbol{X}) = \varrho(\boldsymbol{x}) \det \mathbf{F}(\boldsymbol{X}),$$
(4.10)

$$\mathbf{P}(\boldsymbol{X}) = \det \mathbf{F}(\boldsymbol{X}) \, \mathbf{T}(\boldsymbol{x}) \, \mathbf{F}(\boldsymbol{X})^{-T}$$

$$= \mathbf{T}(\boldsymbol{x}) \operatorname{Cof} \mathbf{F}(\boldsymbol{X}) \qquad \text{(by the Piola transformation)},$$

where the dependence on t has been suppressed. The tensor $\mathbf{P}(\boldsymbol{X})$ is called the *First Piola–Kirchhoff Stress Tensor*. Note that $\mathbf{P}$ is not symmetric; however, $\mathbf{P}\mathbf{F}^T = \mathbf{F}\mathbf{P}^T$ since $\mathbf{T}$ is symmetric. Recalling our earlier proof, we have

$$
\frac{\partial}{\partial X_j} P_{ij} = \frac{\partial}{\partial X_i}(\mathbf{T}\operatorname{Cof}(\mathbf{F})) = \det \mathbf{F}\,\frac{\partial}{\partial X_\ell}T_{ik}\cdot\frac{\partial x_\ell}{\partial X_j}\cdot F_{kj}^{-T}
$$

$$
+\,T_{ik}\frac{\partial}{\partial X_j}((\operatorname{Cof}\mathbf{F})_{kj})^{\;0}
$$

$$
= \det \mathbf{F}\,\frac{\partial}{\partial x_k}T_{ik},
$$

i.e.,

$$
\boxed{\operatorname{Div}\mathbf{P} = \det \mathbf{F}\ \operatorname{div}\mathbf{T},}\tag{4.11}
$$

so

$$
\operatorname{div}\mathbf{T}(\boldsymbol{x}) + \boldsymbol{f}(\boldsymbol{x}) = \frac{1}{\det \mathbf{F}}(\operatorname{Div}\mathbf{P}+\boldsymbol{f}_0)(\boldsymbol{X})
$$

$$
= \varrho\frac{d\boldsymbol{v}}{dt}(\boldsymbol{x}) = \varrho_0\frac{\partial^2 \boldsymbol{u}}{\partial t^2}(\boldsymbol{X})\frac{1}{\det \mathbf{F}}.
$$

Thus, the equations of motion (linear and angular momentum) referred to the reference configuration are

$$
\boxed{\begin{aligned}
\operatorname{Div}\mathbf{P}(\boldsymbol{X}) + \boldsymbol{f}_0(\boldsymbol{X}) &= \varrho_0(\boldsymbol{X})\ddot{\boldsymbol{u}}(\boldsymbol{X}),\\
\mathbf{P}(\boldsymbol{X})\,\mathbf{F}^T(\boldsymbol{X}) &= \mathbf{F}(\boldsymbol{X})\,\mathbf{P}^T(\boldsymbol{X}).
\end{aligned}}\tag{4.12}
$$

Here the dependence of $\mathbf{P}$, $\mathbf{f}_0$, $\ddot{\boldsymbol{u}}$, and $\mathbf{F}$ on time t is not indicated for simplicity; but note that ϱ_0 is independent of t. Alternatively, these equations can be written in terms of a symmetric tensor:

$$
\mathbf{S}(\boldsymbol{X}) = \mathbf{S}(\boldsymbol{X},t) = \mathbf{F}(\boldsymbol{X})^{-1}\mathbf{P}(\boldsymbol{X}) = \det \mathbf{F}\,\mathbf{F}^{-1}\mathbf{T}\mathbf{F}^{-T} = \mathbf{F}^{-1}\mathbf{T}\operatorname{Cof}\mathbf{F}.\tag{4.13}
$$

The tensor $\mathbf{S}$ is the so-called *Second Piola–Kirchhoff Stress Tensor.* Then

$$
\boxed{\begin{aligned}
\operatorname{Div}\mathbf{F}(\boldsymbol{X})\mathbf{S}(\boldsymbol{X}) + \boldsymbol{f}_0(\boldsymbol{X}) &= \varrho_0(\boldsymbol{X})\ddot{\boldsymbol{u}}(\boldsymbol{X}),\\
\mathbf{S}(\boldsymbol{X}) &= \mathbf{S}(\boldsymbol{X})^T.
\end{aligned}}\tag{4.14}
$$

In summary, we have

Cauchy Stress:
$$\mathbf{T} = (\det \mathbf{F})^{-1}\, \mathbf{P}\, \mathbf{F}^T$$
$$= (\det \mathbf{F})^{-1}\, \mathbf{F}\, \mathbf{S}\, \mathbf{F}^T,$$

First Piola–Kirchhoff Stress:
$$\mathbf{P} = (\det \mathbf{F})\, \mathbf{T}\, \mathbf{F}^{-T}$$
$$= \mathbf{F}\, \mathbf{S},$$

Second Piola–Kirchhoff Stress:
$$\mathbf{S} = (\det \mathbf{F})\, \mathbf{F}^{-1}\, \mathbf{T}\, \mathbf{F}^{-T}$$
$$= \mathbf{F}^{-1}\, \mathbf{P}.$$

$$(4.15)$$

4.4 Power

A fundamental property of a body in motion subjected to forces is *power*, the work per unit time developed by the forces acting on the body. Work is "force times distance" and power is "force times velocity." In continuum mechanics, the work per unit time – the power – is the function $\mathcal{P} = \mathcal{P}(t)$ defined by

$$\mathcal{P} = \int_{\Omega_t} \mathbf{f} \cdot \mathbf{v}\, dx + \int_{\partial\Omega_t} \boldsymbol{\sigma}(\boldsymbol{n}) \cdot \mathbf{v}\, dA, \qquad (4.16)$$

where $\boldsymbol{v}$ is the velocity. Since

$$\int_{\partial\Omega_t} \boldsymbol{\sigma}(\boldsymbol{n}) \cdot \boldsymbol{v}\, dA = \int_{\partial\Omega_t} \mathbf{T}\boldsymbol{n} \cdot \boldsymbol{v}\, dA = \int_{\Omega_t} (\boldsymbol{v} \cdot \operatorname{div} \mathbf{T} + \mathbf{T} : \operatorname{grad} \boldsymbol{v})\, dx,$$

we have

$$\mathcal{P} = \int_{\Omega_t} \boldsymbol{v} \cdot (\operatorname{div} \mathbf{T} + \mathbf{f})\, dx + \int_{\Omega_t} \mathbf{T} : \mathbf{D}\, dx$$
$$= \int_{\Omega_t} \boldsymbol{v} \cdot \varrho \frac{d\boldsymbol{v}}{dt}\, dx + \int_{\Omega_t} \mathbf{T} : \mathbf{D}\, dx$$
$$= \frac{d}{dt}\left(\frac{1}{2}\int_{\Omega_t} \varrho \boldsymbol{v} \cdot \boldsymbol{v}\, dx\right) + \int_{\Omega_t} \mathbf{T} : \mathbf{D}\, dx$$
$$= \frac{d\kappa}{dt} + \int_{\Omega_t} \mathbf{T} : \mathbf{D}\, dx, \qquad (4.17)$$

where κ is the kinetic energy,

$$\kappa = \frac{1}{2} \int_{\Omega_t} \varrho v \cdot v \, dx, \tag{4.18}$$

and

$$\mathbf{D} = (\operatorname{grad} v)_{sym} = \frac{1}{2}(\operatorname{grad} v + \operatorname{grad} v^T).$$

Note that $\mathbf{T} : \operatorname{grad} v = \mathbf{T} : (\mathbf{D} + \mathbf{W}) = \mathbf{T} : \mathbf{D}$. The quantity $\mathbf{T} : \mathbf{D}$ is called the *stress power*.

In summary, the total power of a body $\mathcal{B}$ in motion is the sum of the time rate of change of the kinetic energy and the total stress power:

$$\boxed{\mathcal{P} = \frac{d\kappa}{dt} + \int_{\Omega_t} \mathbf{T} : \mathbf{D} \, dx.} \tag{4.19}$$

Equivalently,

$$\boxed{\int_{\Omega_t} \mathbf{f} \cdot v \, dx + \int_{\partial\Omega_t} \mathbf{T}n \cdot v \, dA = \frac{d\kappa}{dt} + \int_{\Omega_t} \mathbf{T} : \mathbf{D} \, dx.} \tag{4.20}$$

In terms of quantities referred to the reference configuration, we have

$$\int_{\Omega_0} \mathbf{f}_0 \cdot \dot{u} \, dX + \int_{\partial\Omega_0} \mathbf{P}n_0 \cdot \dot{u} \, dA_0 = \int_{\Omega_0} \mathbf{P} : \dot{\mathbf{F}} \, dX + \frac{d}{dt}\frac{1}{2} \int_{\Omega_0} \varrho_0 \dot{u} \cdot \dot{u} \, dX.$$

In terms of the second Piola–Kirchhoff stress tensor, the stress power is

$$\int_{\Omega_t} \mathbf{T} : \mathbf{D} \, dx = \int_{\Omega_0} \mathbf{T} : \operatorname{grad} v \det \mathbf{F} \, dX.$$

But

$$
\begin{aligned}
\mathbf{T} : \operatorname{grad} v \det \mathbf{F} &= T_{ij} \frac{\partial}{\partial X_k}\left(\frac{\partial \varphi_i}{\partial t}\right) \cdot \frac{\partial X_k}{\partial x_j} \det \mathbf{F} \\
&= T_{ij} \dot{F}_{ik} F_{kj}^{-1} \det \mathbf{F} \\
&= \dot{F}_{ik}(T_{ij} \det \mathbf{F} F_{kj}^{-1}) \\
&= \dot{F}_{ik} P_{ik} \\
&= \mathbf{F}^T \dot{\mathbf{F}} : \mathbf{S}
\end{aligned}
$$

and

$$\mathbf{F}^T \dot{\mathbf{F}} : \mathbf{S} = \left[\frac{1}{2}(\mathbf{F}^T \dot{\mathbf{F}} + \dot{\mathbf{F}}^T \mathbf{F}) + \frac{1}{2}(\mathbf{F}^T \dot{\mathbf{F}} - \dot{\mathbf{F}}^T \mathbf{F}) \right] \mathbf{S}$$
$$= \dot{\mathbf{E}} : \mathbf{S} + \mathbf{0} = \mathbf{S} : \dot{\mathbf{E}}.$$

Thus

$$\int_{\Omega_t} \mathbf{T} : \mathbf{D} \, dx = \int_{\Omega_0} \mathbf{S} : \dot{\mathbf{E}} \, dX, \qquad (4.21)$$

and, finally,

$$\int_{\Omega_0} \mathbf{f}_0 \cdot \dot{\boldsymbol{u}} \, dX + \int_{\partial\Omega_0} \mathbf{F} \mathbf{S} \, \boldsymbol{n}_0 \cdot \dot{\boldsymbol{u}} \, dA_0 = \int_{\Omega_0} \mathbf{S} : \dot{\mathbf{E}} \, dX$$
$$+ \frac{d}{dt} \left(\frac{1}{2} \int_{\Omega_0} \varrho_0 \dot{\boldsymbol{u}} \cdot \dot{\boldsymbol{u}} \, dX \right).$$

$$(4.22)$$

CHAPTER 5

THE PRINCIPLE OF CONSERVATION OF ENERGY

5.1 Energy and the Conservation of Energy

Energy is a quality of a physical system (e.g., a deformable body) measuring its capacity to do work: A change in energy causes work to be done by the forces acting on the system. A change in energy in time produces a rate of work—i.e., power. So the rate of change of the total energy of a body due to "mechanical processes" (those without a change in temperature) is equal to the mechanical power developed by the forces on the body due to its motion. The total energy is the sum of the kinetic energy (the energy due to motion) κ and the internal energy U due to "everything else" (the deformation, temperature, etc.):

$$\text{Total energy } = \kappa + U.$$

The internal energy depends on the deformation, temperature gradient, and other physical entities. The precise form of this dependency varies from material to material and depends upon the physical "constitution" of the body. In continuum mechanics, it is assumed that a *specific*

An Introduction to Mathematical Modeling: A Course in Mechanics, First Edition. By J. Tinsley Oden
© 2011 John Wiley & Sons, Inc. Published 2011 by John Wiley & Sons, Inc.

internal energy density (energy density per unit mass) exists so that

$$U = \int_{\Omega_0} \varrho_0 e_0(\boldsymbol{X}, t)\, dX = \int_{\Omega_t} \varrho e(\boldsymbol{x}, t)\, dx.$$

Recall that the kinetic energy is given by

$$\kappa = \frac{1}{2} \int_{\Omega_0} \varrho_0\, \dot{\boldsymbol{u}} \cdot \dot{\boldsymbol{u}}\, dX = \frac{1}{2} \int_{\Omega_t} \varrho \boldsymbol{v} \cdot \boldsymbol{v}\, dx$$

and that the power $\mathcal{P}$ is

$$\mathcal{P} = \frac{d\kappa}{dt} + \int_{\Omega_t} \mathbf{T} : \mathbf{D}\, dx = \frac{d\kappa}{dt} + \int_{\Omega_0} \mathbf{S} : \dot{\mathbf{E}}\, dX. \tag{5.1}$$

The change in unit time of the total energy (the time rate of change of $\kappa + U$) produces power and *heating* of the body. The heating per unit time $\dot{Q}$ is of the form

$$\dot{Q} = \int_{\partial\Omega_t} -\boldsymbol{q} \cdot \boldsymbol{n}\, dA + \int_{\Omega_t} r\, dx = \int_{\partial\Omega_0} -\boldsymbol{q}_0 \cdot \boldsymbol{n}_0\, dA_0 + \int_{\Omega_0} r_0\, dX, \tag{5.2}$$

where $\boldsymbol{q}$ is the heat flux entering the body across the surface $\partial\Omega_t$ (minus $\boldsymbol{q}$ indicates heat entering and not leaving the body), r is the heat per unit volume generated by internal sources (e.g., chemical reactions), and $\boldsymbol{q}_0$ and r_0 are their counterparts referred to the reference configurations:

$$\boldsymbol{q} \cdot \det \mathbf{F}\, \mathbf{F}^{-T} \boldsymbol{n}_0 = \boldsymbol{q} \cdot \operatorname{Cof} \mathbf{F} \boldsymbol{n}_0 = (\operatorname{Cof} \mathbf{F})^T \boldsymbol{q} \cdot \boldsymbol{n}_0 = \boldsymbol{q}_0 \cdot \boldsymbol{n}_0,$$
$$r \det \mathbf{F} = r_0.$$

We now arrive at

> **The Principle of Conservation of Energy** *The time rate of change of the total energy is balanced by (equals) the power plus the heating of the body:*
>
> $$\frac{d}{dt}(\kappa + U) = \mathcal{P} + \dot{Q}. \tag{5.3}$$

The principle of conservation of energy is also referred to as the *first law of thermodynamics*. We will encounter it again in Part III.

5.2 Local Forms of the Principle of Conservation of Energy

Recalling the definitions of $\kappa, U, \mathcal{P}$, and $\dot{Q}$, we have

$$
\frac{d}{dt}\left(\frac{1}{2}\int_{\Omega_t}\varrho\boldsymbol{v}\cdot\boldsymbol{v}\,dx+\int_{\Omega_t}\varrho e\,dx\right)=\int_{\Omega_t}\mathbf{T}:\mathbf{D}\,dx+\frac{d\kappa}{dt}
$$
$$
-\int_{\partial\Omega_t}\boldsymbol{q}\cdot\boldsymbol{n}\,dA+\int_{\Omega_t}r\,dx.
$$

Thus

$$
\int_{\Omega_t}\left(\varrho\frac{de}{dt}-\mathbf{T}:\mathbf{D}+\operatorname{div}\boldsymbol{q}-r\right)dx=0,
$$

or

$$
\boxed{\varrho\frac{de}{dt}=\mathbf{T}:\mathbf{D}-\operatorname{div}\boldsymbol{q}+r.} \tag{5.4}
$$

For equivalent results referred to the reference configuration Ω_0, we have

$$
\frac{d}{dt}\left(\kappa+\int_{\Omega_0}\varrho_0 e_0\,dX\right)=\int_{\Omega_0}\mathbf{S}:\dot{\mathbf{E}}\,dX+\frac{d\kappa}{dt}
$$
$$
-\int_{\partial\Omega_0}\boldsymbol{q}_0\cdot\boldsymbol{n}_0\,dA_0+\int_{\Omega_0}r_0\,dX,
$$

or

$$
\int_{\Omega_0}\left(\varrho_0\dot{e}_0-\mathbf{S}:\dot{\mathbf{E}}+\operatorname{Div}\boldsymbol{q}_0-r_0\right)dX=0.
$$

Thus, locally, we have

$$
\boxed{\varrho_0\dot{e}_0=\mathbf{S}:\dot{\mathbf{E}}-\operatorname{Div}\boldsymbol{q}_0+r_0.} \tag{5.5}
$$

CHAPTER 6

THERMODYNAMICS OF CONTINUA AND THE SECOND LAW

In contemplating the thermal and mechanical behavior of the physical universe, it is convenient to think of thermomechanical systems as some open region S of three-dimensional Euclidean space containing, perhaps, one or more deformable bodies. Such a system is *closed* if it does not exchange matter with the complement of S, called the *exterior* of S. There can be an exchange of energy between S and its exterior due to work of external forces and heating (cooling) of S by the transfer of heat from S to its exterior. A system is a *thermodynamic system* if the only exchange of energy with its exterior is a possible exchange of heat and of work done by body and contact forces acting on S. The *thermodynamic state* of a system S is characterized by the values of so-called *thermodynamic state variables*, such as temperature, mass density, etc., which reflect the mechanical and thermal condition of the system; we may write $\mathcal{T}(S, t)$ for the thermodynamic state of system S at time t. If the thermodynamic system does not evolve in time, it is in *thermodynamic equilibrium*. The transition from one state to another is a *thermodynamic process*.

An Introduction to Mathematical Modeling: A Course in Mechanics, First Edition. By J. Tinsley Oden
© 2011 John Wiley & Sons, Inc. Published 2011 by John Wiley & Sons, Inc.

Thus, we may think of the thermodynamic state of a system as the values of certain fields that provide all of the information needed to characterize the system: stress, strain, velocity, etc., and quantities that measure the hotness or coldness of the system and possible rates of change of these quantities. The *absolute temperature* $\theta \in \mathbb{R}$, $\theta > 0$ provides a measure of the hotness of a system and a characterization of the thermal state of a system.[1] Two closed systems $\mathcal{S}_1$ and $\mathcal{S}_2$ are in thermal equilibrium with each other (and with a third system $\mathcal{S}_3$) if they share the same value of θ. Thus, θ is a state variable. The second quantity needed to define the thermodynamic state of a system is the *entropy*, which represents a bound on the amount of heating a system can receive at a given temperature θ.

In continuum mechanics, the total entropy S of a body is the integral of a specific entropy density η (entropy density per unit mass):

$$S = \int_{\Omega_t} \varrho\eta(\boldsymbol{x}, t) \, dx.$$

In classical thermodynamics, the change in entropy between two states $\mathcal{T}(\mathcal{S}, t_1)$ and $\mathcal{T}(\mathcal{S}, t_2)$ of a system measures the quantity of heat received per unit temperature. When $\theta = $ constant, the classical condition is

$$S(\mathcal{S}, t_2) - S(\mathcal{S}, t_1) - \frac{\dot{Q}}{\theta} = 0$$

for reversible processes, and

$$S(\mathcal{S}, t_2) - S(\mathcal{S}, t_1) - \frac{\dot{Q}}{\theta} \geq 0$$

for all possible processes, where $S(\mathcal{S}, t)$ is the total entropy of system $\mathcal{S}$ at time t and $\dot{Q}$ is the heating.

In continuum mechanics, we analogously require

$$\frac{dS}{dt} + \int_{\partial\Omega_t} \frac{1}{\theta} \boldsymbol{q} \cdot \boldsymbol{n} \, dA - \int_{\Omega_t} \frac{r}{\theta} \, dx \geq 0. \tag{6.1}$$

---

[1] The notion of the entropy of a thermodynamic system and the role of the absolute temperature is made more transparent in our treatment of statistical mechanics in Part III.

This is called the *second law of thermodynamics*: The total entropy production per unit time is always ≥ 0. Locally, we have

$$\int_{\Omega_t} \left(\varrho \frac{d\eta}{dt} + \operatorname{div} \frac{\boldsymbol{q}}{\theta} - \frac{r}{\theta} \right) dx \geq 0,$$

or

$$\varrho \frac{d\eta}{dt} + \operatorname{div} \frac{\boldsymbol{q}}{\theta} - \frac{r}{\theta} \geq 0. \tag{6.2}$$

In the material description, we have

$$\varrho_0 \dot{\eta}_0 + \operatorname{Div} \frac{\boldsymbol{q}_0}{\theta} - \frac{r_0}{\theta} \geq 0. \tag{6.3}$$

This relation is called the *Clausius–Duhem inequality*.

We remark that a more general form of (6.1) can be obtained by assuming that there exist an entropy flux $\boldsymbol{h}$ and an entropy supply s such that

$$\frac{dS}{dt} - \int_{\Omega_t} s \, dx + \int_{\partial\Omega_t} \boldsymbol{h} \cdot \boldsymbol{n} \, dA \geq 0, \tag{6.4}$$

or, in place of (6.2),

$$\varrho \frac{d\eta}{dt} + \operatorname{div} \boldsymbol{h} - s \geq 0. \tag{6.5}$$

This, of course, collapses to (6.2) whenever $\boldsymbol{h} = \boldsymbol{q}/\theta$ and $s = r/\theta$.

Table 6.1: Summary of the local conservation equations

Material (LAGRANGIAN)	**Spatial** (EULERIAN)
Conservation of Mass	
$\varrho_0 = \varrho \det \mathbf{F}$	$\dfrac{\partial \varrho}{\partial t} + \mathrm{div}(\varrho \boldsymbol{v}) = 0$
Conservation (Balance) of Linear Momentum	
$\mathrm{Div}\,\mathbf{FS} + \mathbf{f}_0 = \varrho_0 \partial^2 \boldsymbol{u}/\partial t^2$	$\mathrm{div}\,\mathbf{T} + \mathbf{f} = \varrho \left(\dfrac{\partial \boldsymbol{v}}{\partial t} + \boldsymbol{v} \cdot \mathrm{grad}\,\boldsymbol{v} \right)$
Conservation (Balance) of Angular Momentum	
$\mathbf{S} = \mathbf{S}^T$	$\mathbf{T} = \mathbf{T}^T$
Conservation of Energy	
$\varrho_0 \dot{e}_0 = \mathbf{S} : \dot{\mathbf{E}} - \mathrm{Div}\,\boldsymbol{q}_0 + r_0$	$\varrho \dfrac{\partial e}{\partial t} + \varrho \boldsymbol{v} \cdot \mathrm{grad}\,e = \mathbf{T} : \mathbf{D} - \mathrm{div}\,\boldsymbol{q} + r$
Second Law of Thermodynamics (The Clausius–Duhem Inequality)	
$\varrho_0 \dot{\eta}_0 + \mathrm{Div}\,\dfrac{\boldsymbol{q}_0}{\theta} - \dfrac{r_0}{\theta} \geq 0$	$\varrho \dfrac{\partial \eta}{\partial t} + \varrho \boldsymbol{v} \cdot \mathrm{grad}\,\eta + \mathrm{div}\,\dfrac{\boldsymbol{q}}{\theta} - \dfrac{r}{\theta} \geq 0$

CHAPTER 7

CONSTITUTIVE EQUATIONS

Given initial data (i.e. given all of the information needed to describe the body $\mathcal{B}$ and its thermodynamic state $\mathcal{S}$ while it occupies its reference configuration at time $t = 0$),

$$\left(\Omega_0,\ \partial\Omega_0,\ \mathbf{f}_0(\boldsymbol{X}, t),\ t \in [0, T],\ \boldsymbol{v}_0 = \dot{\boldsymbol{u}}(\boldsymbol{X}, 0),\ \theta_0(\boldsymbol{X}) = \theta(\boldsymbol{X}, 0), \dots\right),$$

we wish to use the equations of continuum mechanics (conservation of mass, energy, balance of linear and angular momentum, the second law of thermodynamics) to determine the behavior of the body (the motion, deformation, temperature, stress, heat flux, entropy, ...) at each $\boldsymbol{X} \in \overline{\Omega_0}$ for any time t in some interval $[0, T]$.

Unfortunately, we do not have enough information to solve this problem. The balance laws apply to all materials, and we know that different materials respond to the same stimuli in different ways depending on their *constitution*. To complete the problem (to "close" the system of equations), we must supplement the basic equations with *constitutive equations* that characterize the material(s) of which the body is composed. Thus, the constitutive equations impose constraints on the possible responses of the material body that describe how the particular material behaves when subjected to various stimuli: forces, heating, etc.

In recognizing that additional relations are needed to close the system, we ask what variables can we identify as natural "dependent variables"

An Introduction to Mathematical Modeling: A Course in Mechanics, First Edition. By J. Tinsley Oden and which are "independent." Our choice is to choose as primitive (state) variables those features of the response naturally experienced by observation – by our physical senses: the motion (or displacement), the rate of motion, the deformation or rate of deformation, the hotness or coldness (the temperature) or its gradients, etc., or the "histories" of these quantities. If we then knew how, for example, the stress, heat flux, internal energy, and entropy were dependent on the state variables *for a given material*, we would hope to have sufficient information to completely characterize the behavior of the body under the action of given stimuli (loads, heating, etc.). Thus, we seek constitutive equations of the type

$$
\begin{aligned}
\mathbf{T} &= \mathcal{T}\left(\boldsymbol{x},t,\wedge\right), \\
q &= \mathbf{Q}\left(\boldsymbol{x},t,\wedge\right), \\
e &= \xi\left(\boldsymbol{x},t,\wedge\right), \\
\eta &= \mathcal{H}\left(\boldsymbol{x},t,\wedge\right),
\end{aligned}
\tag{7.1}
$$

where, for the moment, $\wedge$ denotes everything we might expect to influence the stress, heat flux, internal energy, and entropy at a point $\boldsymbol{x} \in \Omega_t$ at time t (for example, $\wedge = (\mathbf{F}, \mathbf{C}, \theta, \nabla\theta, \mathbf{L}, \mathbf{D}, \ldots)$). The functions $\mathcal{T}$, $\mathbf{Q}, \mathcal{E}, \mathcal{H}$ defining the constitutive equations are called *response functions* (or, response functionals).

7.1 Rules and Principles for Constitutive Equations

What rules-of-thumb or even fundamental principles govern the forms of the response functions for real materials? We list some of the most important rules and principles:

---

1. The Principle of Determinism *The behavior of a material at a point $\mathbf{X}$ occupying $\boldsymbol{x}$ at time t is determined by the history of the primitive variables $\wedge$.*

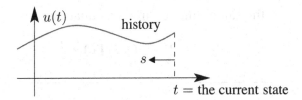

Figure 7.1: Time history.

In other words, the conditions that prevail at (x, t) depend upon the past behavior of the array $\wedge$ (not the future). In general, a "history" of a function $u = u(x, t)$, up to current time t, is the set

$$u^t(s) = \{u(x, t - s), \ s \geq 0\}.$$

The history from the reference configuration is then $u(x, t-s), t \geq s \geq 0$ (see Fig. 7.1). Then we should, in general, replace $\wedge$ by $\wedge^t(s)$ in the constitutive equations unless conditions prevail (in the constitution of the material) that suggest the response depends only on values at the current time.

---

2. The Principle of Material Frame Indifference *The form of the constitutive equations must be invariant under changes of the frame of reference: They must be independent of the observer.*

For example, if we rotate the coordinates x into a system x^* (and translate the origin)

$$\left.\begin{array}{l} x^* = \mathbf{Q}(t)x + c(t) \\ \mathbf{Q}^T(t) = \mathbf{Q}^{-1}(t) \end{array}\right\} \Rightarrow \text{"Change of the Observer"} \qquad (7.2)$$

and change the clock: $t^* = t - a$, the constitutive functions should remain invariant. For example,

$$\text{if } \mathbf{T} = \mathcal{T}(x, t), \quad \text{then } \mathbf{T}^* = \mathcal{T}(x^*, t^*),$$

where $\mathcal{T}$ is the same function in both equations and

$$\mathbf{T}^* = \mathbf{Q}(t)\mathbf{T}\mathbf{Q}(t)^T. \tag{7.3}$$

---

3. The Principle of Physical Consistency *The constitutive equations cannot violate or contradict the physical principles of mechanics (the conservation of mass, energy , balance of momentum, or the Clausius–Duhem inequality).*

---

4. The Principle of Material Symmetry *For every material, there is a group $\mathcal{G}$ of unimodular transformations of the material coordinates, called the isotropy group of the material, under which the forms of the constitutive functions remain invariant.*

- The unimodular tensors H are those for which $\det H = \pm 1$.

- The group $\mathcal{G}$ is a group of unimodular linear transformations with group operation = composition (matrix multiplication). Thus,

    1. $\mathbf{A}, \mathbf{B} \in \mathcal{G} \Rightarrow \mathbf{AB} \in \mathcal{G}$;
    2. $\mathbf{A}(\mathbf{BC}) = (\mathbf{AB})\mathbf{C}$;
    3. $\exists \mathbf{1} \in \mathcal{G}$ such that $\mathbf{A} \cdot \mathbf{1} = \mathbf{A}$;
    4. $\forall \mathbf{A} \in \mathcal{G}, \exists \mathbf{A}^{-1} \in \mathcal{G}$, such that $\mathbf{A}\,\mathbf{A}^{-1} = \mathbf{1}$.

- Let $\mathcal{O}^+$ be the group of rotations (called the proper orthogonal group). If, at a material point $\mathbf{X}$, the symmetry group $\mathcal{G}$ equals $\mathcal{O}^+$, then the material is *isotropic* at $\bar{\mathbf{X}}$; otherwise, it is *anisotropic*.

> **5. The Principle of Local Action** *The dependent (primitive) consti-*
> *tutive variables at* X *are not affected by the actions of independent*
> *variables* $((\mathbf{T}, \mathbf{q}, e, \eta)$ *for example) at points distant from* X.

There are other rules. For example:

- **Dimensional consistency:** Terms in the constitutive equations must, of course, be dimensionally consistent (this can be interpreted as a corollary to Principle 3). Beyond this, dimensional analysis can be used to extract information on the constitutive variables.

- **Existence, well-posedness:** The constitutive equations must be such that there exist solutions to properly posed boundary and initial-value problems resulting from use of the equations of continuum mechanics.

- **Equipresence:** When beginning to characterize the response functions of a material, be sure that the dependent variables in $\wedge$ are "equally present" – in other words, until evidence from some other source suggests otherwise, assume that all the constitutive functions for $\mathbf{T}, \mathbf{q}, e, \eta$ depend on the same full list of variables $\wedge$.

7.2 Principle of Material Frame Indifference

7.2.1 Solids

Consider a "change in the observer":

$$\mathbf{x}^*(t) = \mathbf{Q}(t)\mathbf{x}(t) + \mathbf{c}(t),$$
$$\mathbf{Q}(t)^{-1} = \mathbf{Q}(t)^T, \tag{7.4}$$

where $\mathbf{x}(t) = \varphi(\mathbf{X}, t)$. Clearly, $\mathbf{F}^* = \nabla \mathbf{x}^*$ and $\mathbf{F} = \nabla \mathbf{x}$, so that

$$\boxed{\mathbf{F}^* = \mathbf{Q}\mathbf{F} \quad \text{and} \quad \det \mathbf{F}^* = \det \mathbf{F}.} \tag{7.5}$$

If $n^* = \mathbf{Q}n$ and $\boldsymbol{\sigma}^*(n^*) = \mathbf{Q}\boldsymbol{\sigma}(n)$, then

$$\mathbf{T}^*n^* = \mathbf{QTQ}^T n^* \quad \Rightarrow \quad \boxed{\mathbf{T}^* = \mathbf{QTQ}^T.} \tag{7.6}$$

Suppose

$$\mathbf{T} = \mathcal{T}(\mathbf{F})$$

(Local Action $\Rightarrow$ $\mathbf{T}$ depends on $\nabla\varphi$, not φ).

Then

$$\mathbf{T}^* = \mathcal{T}(\mathbf{F}^*).$$

Thus, the material response function is independent of the observer, if, $\forall\, \mathbf{Q}(t),\ \mathbf{Q}(t)^T = \mathbf{Q}(t)^{-1}$, we have

$$\boxed{\mathcal{T}(\mathbf{QF}) = \mathbf{Q}\mathcal{T}(\mathbf{F})\mathbf{Q}^T.} \tag{7.7}$$

Example If $\mathcal{T}(\mathbf{F}) = \mathbf{T}$, then $\mathcal{T}(\mathbf{F}) = \mathbf{R}\mathcal{T}(\mathbf{U})\mathbf{R}^T$, since $\mathcal{T}(\mathbf{R}^T\mathbf{F}) = \mathbf{R}\mathcal{T}(\mathbf{R}^T\mathbf{R}\mathbf{U})\mathbf{R}^T = \mathbf{R}\mathcal{T}(\mathbf{U})\mathbf{R}^T$. This shows that we can also express $\mathbf{T}$ as a function of $\mathbf{U}$ or, since $\mathbf{U}^2 = \mathbf{C}$, of $\mathbf{C}$ or $\mathbf{E}$; i.e., $\mathbf{T} = \mathcal{T}(\mathbf{F}) = \mathbf{R}\mathcal{T}(\mathbf{U})\mathbf{R}^T = \mathbf{F}\mathbf{U}^{-1}\mathcal{T}(\mathbf{C}^{1/2})\mathbf{U}^{-1}\mathbf{F}^T = \mathbf{F}\widehat{\mathcal{T}}(\mathbf{C})\mathbf{F}^T$. $\qquad\square$

In general, constitutive equations are said to characterize the response of a material at a point (a material particle) in the body to values of the state variables (deformation, temperature, velocity gradients, ...). The principles discussed up to now lead us to the notion that the response can depend upon the entire history of appropriate state variables. A material is called *thermoelastic* if the constitutive functions for $(\mathbf{T}, q, e, \eta)$ depend only on the present values of $\mathbf{F}$, θ, and $\nabla\theta$ at the point. If all of the material points of the body are composed of the same material, the body is said to be *homogeneous*; otherwise, it is inhomogeneous. We will give examples of constitutive equations for thermoelastic solids in the next chapter.

Let us consider the Helmholtz free energy ψ_0 per unit volume in the reference configuration, and let the heat flux q_0 refer to the reference configuration of a thermoelastic body. As an example, suppose $\psi_0 = \Psi(\mathbf{F}, \theta)$ and $q_0 = \mathbf{Q}(\mathbf{F}, \theta, \mathbf{G})$, where $\mathbf{G} = \nabla\theta$. Let $\mathbf{O}$ denote an orthogonal transformation defining a rigid rotation of the spatial coordinate, as in (7.4). Then the principle of material frame indifference asserts that

$$\psi_0(\mathbf{OF}, \theta) = \psi_0(\mathbf{F}, \theta) \quad \text{and} \quad \mathbf{Q}(\mathbf{OF}, \theta, \mathbf{G}) = \mathbf{Q}(\mathbf{F}, \theta, \mathbf{G}).$$

In particular, if, according to the polar decomposition theorem, $\mathbf{F} = \mathbf{RU}$, replacing $\mathbf{O}$ by $\mathbf{R}$ shows that $\psi_0(\mathbf{F}, \theta) = \psi_0(\mathbf{U}, \theta)$ and $\mathbf{Q}(\mathbf{F}, \theta, \mathbf{G}) = \mathbf{Q}(\mathbf{U}, \theta, \mathbf{G})$.

7.2.2 Fluids

A material is generally called a fluid if any change in thermomechanical deformations from its current configuration are unaffected by changes in the reference configurations due to unimodular transformations. An *ideal fluid* is a material body in which all motions are *isochoric* (i.e., volume preserving) and the density ϱ is constant. Such fluids are said to be *incompressible*. We expect then that $\overline{\det \mathbf{F}} = 0$, but since $\overline{\det \mathbf{F}} = \det \mathbf{F} \operatorname{div} \boldsymbol{v}$, all motions of incompressible fluids satisfy the constraint,

$$\operatorname{div} \boldsymbol{v} = 0.$$

The Cauchy stress tensor in such fluids is simply

$$\mathbf{T} = -p\,\mathbf{I}, \tag{7.8}$$

where $p = p(\boldsymbol{x}, t, \theta)$ is the static pressure, also called the "hydrostatic" pressure. Obviously, this constitutive equation is form-invariant under changes in the spatial frame of reference; i.e., it obeys the principle of material frame indifference.

A more general class of fluids is characterized by constitutive equations of the form (ignoring θ)

$$\mathbf{T} = \mathcal{F}(\mathbf{L}), \tag{7.9}$$

where $\mathcal{F}(0) = -p\,\mathbf{I}$, p being the static pressure. If we also assume incompressibility,

$$\operatorname{tr} \mathbf{L} = \operatorname{div} \boldsymbol{v} = 0,$$

then

$$\mathbf{T} = -p\,\mathbf{I} + \mathcal{F}_0(\mathbf{D}), \tag{7.10}$$

$\mathcal{F}_0(\mathbf{0}) = 0$. If $\mathcal{F}_0$ is linear in $\mathbf{D}$, then a necessary and sufficient condition that $\mathcal{F}_0(\mathbf{D})$ is invariant under a change of the observer is that there exists a constant (or scalar) $\mu = \mu(t) > 0$, such that $\mathcal{F}_0(\mathbf{D}) = 2\mu\mathbf{D}$. Then

$$\boxed{\mathbf{T} = -p\,\mathbf{I} + 2\mu\mathbf{D}.} \tag{7.11}$$

This is the classical constitutive equation for Cauchy stress in a Newtonian fluid (a viscous incompressible fluid), where μ is the *viscosity of the fluid*. The most important example of fluids often modeled as Newtonian fluids is water.

There are, of course, much more general characterizations of fluids. The equation

$$\mathbf{T} = -p\,\mathbf{I} + f_1\,\mathbf{D} + f_2\,\mathbf{D}^2 + f_3\,\boldsymbol{g} \otimes \boldsymbol{g} + f_4\,(\boldsymbol{g} \otimes \mathbf{D}\boldsymbol{g}) + f_5\,(\boldsymbol{g} \otimes \mathbf{D}^2\boldsymbol{g}), \tag{7.12}$$

where $\boldsymbol{g} = \nabla\theta$, characterizes a class of non-Newtonian fluids. The coefficients $f_1, f_2, \ldots, f_5$ in (7.12) are understood to be functions of invariants of $\mathbf{D}$ and $\boldsymbol{g}$, $\mathbf{D}^2$, etc. Examples of fluids often modeled as non-Newtonian fluids are blood, syrup, and paint.

7.3 The Coleman–Noll Method: Consistency with the Second Law of Thermodynamics

The most important application of the principle of physical consistency is afforded by the Coleman–Noll argument, which places restrictions on the nature of constitutive equations imposed by the second law of thermodynamics. The first step is to introduce the *Helmholtz free energy* ψ per unit mass, defined by

$$\psi = e - \theta\eta, \tag{7.13}$$

where e is the internal energy density, η the entropy density, and θ the absolute temperature. Instead of (6.2), we now have

$$-\varrho\frac{d\psi}{dt} - \varrho\eta\frac{d\theta}{dt} + \mathbf{T} : \mathbf{D} - \frac{\boldsymbol{q}}{\theta} \cdot \operatorname{grad}\theta \geq 0, \tag{7.14}$$

where we also introduced (5.4). In the reference configuration this takes on the form

$$-\varrho_0 \dot{\psi}_0 - \varrho_0 \eta_0 \dot{\theta} + \mathbf{S} : \dot{\mathbf{E}} - \frac{1}{\theta} \boldsymbol{q}_0 \cdot \nabla \theta \geq 0. \qquad (7.15)$$

Now let us suppose that, for example, the constitutive equation for the Helmholtz free energy takes on the form

$$\psi_0 = \Psi(\mathbf{E}, \theta, \nabla\theta). \qquad (7.16)$$

Then,

$$\dot{\psi}_0 = \frac{\partial \Psi}{\partial \mathbf{E}} : \dot{\mathbf{E}} + \frac{\partial \Psi}{\partial \theta}\dot{\theta} + \frac{\partial \Psi}{\partial \nabla\theta} \cdot \nabla\dot{\theta}. \qquad (7.17)$$

Therefore, (7.15) can be written

$$\left(\mathbf{S} - \varrho_0 \frac{\partial \Psi}{\partial \mathbf{E}}\right) : \dot{\mathbf{E}} - \varrho_0 \left(\frac{\partial \Psi}{\partial \theta} + \eta_0\right)\dot{\theta} - \varrho_0 \frac{\partial \Psi}{\partial \nabla\theta} \cdot \nabla\dot{\theta} - \frac{1}{\theta}\boldsymbol{q}_0 \cdot \nabla\theta \geq 0.$$
$$(7.18)$$

We now apply the argument behind the Coleman–Noll method: Since (7.18) must hold for rates $(\dot{\mathbf{E}}, \dot{\theta}, \nabla\dot{\theta})$ of arbitrary sign, it is sufficient that the coefficients of the rates vanish, so we have

$$\mathbf{S} = \varrho_0 \frac{\partial \Psi}{\partial \mathbf{E}}, \quad \eta_0 = -\frac{\partial \Psi}{\partial \theta}, \quad \frac{\partial \Psi}{\partial \nabla\theta} = 0, \qquad (7.19)$$

and

$$-\frac{1}{\theta}\boldsymbol{q}_0 \cdot \nabla\theta \geq 0. \qquad (7.20)$$

This is an important result. In order that the constitutive equation for ψ_0 and $\boldsymbol{q}_0$ be "physically consistent," it is sufficient that equations (7.19) and inequality (7.20) hold. We thus obtain constitutive equations for $\mathbf{S}$ and η_0 directly from that for ψ_0. The free energy cannot depend on the gradient of the temperature, $\nabla\theta$. Likewise, "heat must flow from hot to cold," a condition captured by inequality (7.20).

We observe that the free energy determines the conservative or non-dissipative part of the stress. For example, if

$$\mathbf{S} = \mathcal{F}(\mathbf{E}) + \mathcal{I}(\dot{\mathbf{E}}), \tag{7.21}$$

where $\mathcal{I}(\dot{\mathbf{E}})$ denotes a "dissipative" stress, then the first term in (7.18) becomes

$$\left(\mathcal{F}(\mathbf{E}) + \mathcal{I}(\dot{\mathbf{E}}) - \varrho_0 \frac{\partial \Psi}{\partial \mathbf{E}} \right) : \dot{\mathbf{E}}$$

and the Coleman–Noll method requires that

$$\mathcal{F}(\mathbf{E}) = \varrho_0 \frac{\partial \Psi}{\partial \mathbf{E}} \quad \text{and} \quad \mathcal{I}(\dot{\mathbf{E}}) : \dot{\mathbf{E}} - \frac{1}{\theta} \mathbf{q}_0 \cdot \nabla \theta \geq 0. \tag{7.22}$$

This result is also captured by the notion of the *internal dissipation field* δ_0, defined by

$$\delta_0 = \mathbf{S} : \dot{\mathbf{E}} - \varrho_0 \left(\dot{\psi}_0 + \eta_0 \dot{\theta} \right) = \mathrm{Div}\mathbf{q}_0 - r_0 + \varrho_0 \theta \dot{\eta}_0. \tag{7.23}$$

The Clausius–Duhem inequality thus demands that

$$\delta_0 - \frac{1}{\theta} \mathbf{q}_0 \cdot \nabla \theta \geq 0. \tag{7.24}$$

In the example (7.22), clearly $\delta_0 = \mathcal{I}(\dot{\mathbf{E}}) : \dot{\mathbf{E}}$.

CHAPTER 8

EXAMPLES AND APPLICATIONS

At this point, we have developed all of the components of the general continuum theory of thermomechanical behavior of material bodies. The classical mathematical models of fluid mechanics, heat transfer, and solid mechanics can now be constructed as special cases. In this chapter, we develop several of the more classical mathematical models as special applications of the general theory. Our goal is merely to develop the system of equations that govern classical models; we do not dwell on the construction of solutions or on other details.

8.1 The Navier–Stokes Equations for Incompressible Flow

We begin with one of the most important example in continuum mechanics: The equations governing the flow of a viscous incompressible fluid. These are the celebrated Navier–Stokes equations for incompressible flow. They are used in countless applications to study the flow of fluids, primarily but not exclusively water or even air under certain conditions. We write down the balance laws and the constitutive equation:

An Introduction to Mathematical Modeling: A Course in Mechanics, First Edition. By J. Tinsley Oden
© 2011 John Wiley & Sons, Inc. Published 2011 by John Wiley & Sons, Inc.

Conservation of Mass   (Recall (2.4))

$$\frac{\partial \varrho}{\partial t} + \mathrm{div}(\varrho \boldsymbol{v}) = 0. \tag{8.1}$$

For an incompressible fluid, we have

$$0 = \frac{d}{dt}\int_{\Omega_t} dx = \int_{\Omega_0} \overline{\det \mathbf{F}}\, dX = \int_{\Omega_0} \det \mathbf{F}\, \mathrm{div}\, \boldsymbol{v}\, dX = \int_{\Omega_t} \mathrm{div}\, \boldsymbol{v}\, dx,$$

which implies that

$$\mathrm{div}\, \boldsymbol{v} = 0. \tag{8.2}$$

Thus

$$0 = \frac{\partial \varrho}{\partial t} + \boldsymbol{v} \cdot \mathrm{grad}\, \varrho + \varrho\, \mathrm{div}\, \boldsymbol{v} = \frac{\partial \varrho}{\partial t} + \boldsymbol{v} \cdot \mathrm{grad}\, \varrho = \frac{d\varrho}{dt}, \tag{8.3}$$

which means that ϱ is constant.

Conservation of Momentum   (Recall (4.8))

$$\varrho\frac{\partial \boldsymbol{v}}{\partial t} + \varrho \boldsymbol{v} \cdot \mathrm{grad}\, \boldsymbol{v} - \mathrm{div}\, \mathbf{T} = \mathbf{f}, \quad \forall(\boldsymbol{x}, t) \in \Omega_t \times (0, T),$$
$$\mathbf{T} = \mathbf{T}^T. \tag{8.4}$$

Constitutive Equation   (Recall (7.11))

$$\mathbf{T} = -p\,\mathbf{I} + 2\mu\mathbf{D},$$
$$\mathbf{D} = \frac{1}{2}\left(\mathrm{grad}\, \boldsymbol{v} + (\mathrm{grad}\, \boldsymbol{v})^T\right). \tag{8.5}$$

The Navier–Stokes equations   Finally, introducing (8.5) into (8.4)$_1$ and adding (8.2) gives the celebrated Navier–Stokes equations,

$$\boxed{\begin{aligned} \varrho\frac{\partial \boldsymbol{v}}{\partial t} + \varrho \boldsymbol{v} \cdot \mathrm{grad}\, \boldsymbol{v} - \mu\Delta \boldsymbol{v} + \mathrm{grad}\, p &= \mathbf{f}, \\ \mathrm{div}\, \boldsymbol{v} &= 0, \end{aligned}} \tag{8.6}$$

where Δ = vector Laplacian = div grad.

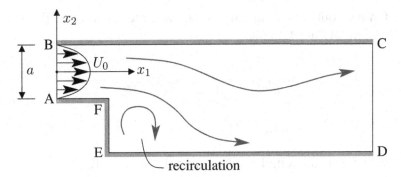

Figure 8.1: Geometry of the backstep channel flow.

Application: An initial-boundary-value problem As a standard model problem, consider the problem of viscous flow through a channel with a "backstep" as shown in Fig. 8.1.
Initial conditions:

$$v(x, 0) = v_0(x), \qquad (8.7)$$

where the initial field must satisfy div $v_0 = 0$, which implies that

$$\int_{\partial \Omega_0} v_0 \cdot n \, dA = 0. \qquad (8.8)$$

Boundary conditions:

1. On segment BC and DE $\cup$ EF $\cup$ FA, we have the no-slip boundary condition $v = 0$.

2. On the "in-flow boundary" AB, we prescribe the Poiseuille flow velocity profile (see Fig. 8.1):

$$v_1(0, x_2, t) = \left(1 - \left(\frac{2x_2}{a}\right)^2\right) U_0,$$

$$v_2(0, x_2, t) = 0. \qquad (8.9)$$

3. On the "out-flow boundary" CD, there are several possibilities. A commonly used one is

$$-p + \mu \left. \frac{\partial v_1}{\partial x_1} \right|_{x_1=L} = \mathbf{T} \mathbf{e}_1 \big|_{x_1=L} = 0,$$

(8.10)

$$v_2(L, x_2, t) = 0,$$

where L is the length of the channel, i.e., $L = |\mathrm{BC}|$.

8.2 Flow of Gases and Compressible Fluids: The Compressible Navier–Stokes Equations

In the case of compressible fluids and gases, the so-called compressible Navier–Stokes equations can be derived from results we laid down earlier. These are the equations governing gas dynamics, aerodynamics, and the mechanics of high-speed flows around objects such as aircraft, missiles, and projectiles.

Typical constitutive equations for stress, heat flux, and entropy are

$$\mathbf{T} = -p\mathbf{I} + \lambda \, \mathrm{tr}(\mathbf{D})\mathbf{I} + 2\mu\mathbf{D} - \alpha\theta\mathbf{I},$$

$$\boldsymbol{q} = -k \, \mathrm{grad} \, \theta,$$

$$\frac{\partial \eta}{\partial \varrho} = \varrho^{-2} p,$$

(8.11)

where p is now the thermodynamic pressure, usually given as a function of θ and ϱ by an "equation of state," λ and μ are bulk and shear viscosities, α a thermal parameter, and k the thermal conductivity. Here θ is usually interpreted as a change in temperature from some reference state. Because $(8.11)_1$ is linear in the deformation rate $\mathbf{D}$, $(8.11)_1$ is said to describe a Newtonian fluid. Relation $(8.11)_2$ is the classical Fourier's Law.

The equations of balance of mass, linear momentum, and energy now assume the forms

$$\frac{\partial \varrho}{\partial t} + \operatorname{div}(\varrho \boldsymbol{v}) = 0,$$

$$\varrho \frac{\partial \boldsymbol{v}}{\partial t} + \varrho \boldsymbol{v} \cdot \operatorname{grad} \boldsymbol{v} = -\operatorname{grad} p + (\lambda + \mu) \operatorname{grad} \operatorname{div} \boldsymbol{v}$$
$$+ \mu \Delta \boldsymbol{v} - \alpha \operatorname{grad} \theta, \tag{8.12}$$

$$\varrho \theta \left(\varrho^{-2} p \dot{\varrho} + \frac{\partial \eta}{\partial \theta} \dot{\theta} \right) = k \Delta \theta + r + \lambda (\operatorname{tr} \mathbf{D})^2 + 2\mu \operatorname{tr}(\mathbf{D}^2).$$

Here $\dot{\varrho} = d\varrho/dt$, $\dot{\theta} = d\theta/dt$, and $\operatorname{tr} \mathbf{D} = \operatorname{div} \boldsymbol{v}$. Typical boundary and initial conditions are of the form

$$\varrho(\boldsymbol{x}, 0) = \varrho_0(\boldsymbol{x}), \quad \boldsymbol{v}(\boldsymbol{x}, 0) = \boldsymbol{v}_0(\boldsymbol{x}), \quad \theta(\boldsymbol{x}, 0) = \theta_0(\boldsymbol{x}), \tag{8.13}$$

for given data, $\varrho_0, \boldsymbol{v}_0, \theta_0$, with $\boldsymbol{x} \in \Omega \subset \mathbb{R}^3$, and

$$\boldsymbol{n} \cdot \left(-p\mathbf{I} + \lambda \operatorname{div} \boldsymbol{v} \, \mathbf{I} + 2\mu \mathbf{D} \right) = \boldsymbol{g}(\boldsymbol{x}, t) \qquad \text{on } \Gamma_1 \times (0, \tau),$$
$$\boldsymbol{v}(\boldsymbol{x}, t) = \hat{\boldsymbol{v}}(\boldsymbol{x}, t) \qquad \text{on } \Gamma_2 \times (0, \tau), \tag{8.14}$$
$$-\boldsymbol{n} \cdot k \nabla \theta = \hat{q} \qquad \text{on } \Gamma_3 \times (0, \tau),$$

where $\boldsymbol{g}$, $\hat{\boldsymbol{v}}$, and $\hat{q}$ are given boundary data and $\boldsymbol{n}$ is a unit extension normal to domain Ω, and Γ_1, Γ_2, and Γ_3 are subsets of the boundary $\partial\Omega$ of Ω.

8.3 Heat Conduction

Ignoring motion and deformation for the moment, consider a rigid body being heated by some outside source. We return to the Helmholtz free energy of (7.13) and the constitutive equations (7.19). The constitutive equations are assumed to be

$$\psi_0 = -\frac{\alpha}{2}(\theta - \theta_0)^2 - \beta(\theta - \theta_0) + \gamma(\theta_0), \tag{8.15}$$

where α and β are positive constants, θ_0 is a given uniform reference temperature, and $\gamma(\theta_0)$ is a constant depending only on θ_0. The heat flux is given by

$$\boldsymbol{q}_0 = -k\nabla(\theta - \theta_0) \qquad \text{(Fourier's Law)}. \qquad (8.16)$$

Then,

$$\eta_0 = -\frac{\partial \psi_0}{\partial \theta} = \alpha(\theta - \theta_0) + \beta, \qquad (8.17)$$

$$\mathbf{S} = \varrho_0 \frac{\partial \psi_0}{\partial \mathbf{E}} = 0 \qquad \text{(irrelevant)}.$$

The specific heat c of a material is the change in internal energy of a unit mass of material due to a unit change in temperature at constant deformation gradient $\mathbf{F}$ and is defined by

$$c = \frac{\partial e_0}{\partial \theta}, \qquad (8.18)$$

where e_0 is the internal energy per unit mass in the reference configuration. Using the definition of ψ_0, we easily show that

$$c = \theta \frac{\partial \eta_0}{\partial \theta}$$
$$= \alpha\theta. \qquad (8.19)$$

Finally, turning to the equation for the conservation of energy (e.g., (5.5)), we have

$$\varrho_0 \dot{e}_0 = \varrho_0 \frac{\partial e_0}{\partial \theta} \dot{\theta}$$
$$= \varrho_0 c\dot{\theta}$$
$$= -\operatorname{Div} \boldsymbol{q}_0 + r_0$$
$$= -\operatorname{Div}(-k\nabla\theta) + r_0$$

or

$$\boxed{\varrho_0 c \frac{\partial \theta}{\partial t} - \nabla \cdot k\nabla\theta = r_0.} \qquad (8.20)$$

This is the classical heat equation. In general, one can consider using the approximate specific heat $c = \alpha\theta \approx \alpha\theta_0$.

8.4   Theory of Elasticity

We consider a deformable body $\mathcal{B}$ under the action of forces (body forces $\mathbf{f}$ and prescribed contact forces $\boldsymbol{\sigma}(n, x) = \boldsymbol{g}(x)$ on $\partial\Omega_t$). The body is constructed of a material which is homogeneous and isotropic and is subjected to only isothermal (θ = constant) and adiabatic ($q = 0$) processes. The sole constitutive equation is

$$\psi = \text{free energy} = \widehat{\Psi}(\mathbf{X}, t, \theta, \mathbf{E}) = \Psi(\mathbf{E}). \tag{8.21}$$

Let $\widetilde{\Psi}$ denote the free energy per unit volume: $\widetilde{\Psi} = \varrho_0\Psi$. So the constitutive equation for stress is

$$\mathbf{S} = \left.\frac{\partial\widetilde{\Psi}(\mathbf{E})}{\partial\mathbf{E}}\right|_{\text{sym}}. \tag{8.22}$$

In this case, the free energy is called the *stored energy function*, or the *strain energy function*. Since $\widetilde{\Psi}(\cdot)$ (and $\partial\psi/\partial\mathbf{E}$) must be form-invariant under changes of the observer and since $\mathcal{B}$ is isotropic, $\widetilde{\Psi}(\mathbf{E})$ must depend on invariants of $\mathbf{E}$:

$$\widetilde{\Psi}(\mathbf{E}) = W(I_E, I\!I_E, I\!I\!I_E). \tag{8.23}$$

Or, since $\mathbf{E} = (\mathbf{C}-I)/2$, we could also write $\widetilde{\Psi}$ as a function of invariants of $\mathbf{C}$:

$$\widetilde{\Psi} = \widehat{W}(I_C, I\!I_C, I\!I\!I_C). \tag{8.24}$$

The constitutive equation for stress is then

$$\boxed{\begin{aligned}\mathbf{S} &= \frac{\partial W}{\partial I_E}\cdot\frac{\partial I_E}{\partial\mathbf{E}} + \frac{\partial W}{\partial I\!I_E}\cdot\frac{\partial I\!I_E}{\partial\mathbf{E}} + \frac{\partial W}{\partial I\!I\!I_E}\cdot\frac{\partial I\!I\!I_E}{\partial\mathbf{E}}\\ &= \frac{\partial\widehat{W}}{\partial I_C}\cdot\frac{\partial I_C}{\partial\mathbf{C}} + \frac{\partial\widehat{W}}{\partial I\!I_C}\cdot\frac{\partial I\!I_C}{\partial\mathbf{C}} + \frac{\partial\widehat{W}}{\partial I\!I\!I_C}\cdot\frac{\partial I\!I\!I_C}{\partial\mathbf{C}}.\end{aligned}} \tag{8.25}$$

and we note that

$$
\begin{aligned}
\frac{\partial I_E}{\partial \mathbf{E}} &= \mathbf{I}, \\
\frac{\partial I\!\!I_E}{\partial \mathbf{E}} &= (\operatorname{tr} \mathbf{E}^{-1})\mathbf{I} - \mathbf{E}^{-T}\operatorname{Cof} \mathbf{E}, \\
\frac{\partial I\!\!I\!\!I_E}{\partial \mathbf{E}} &= \operatorname{Cof} \mathbf{E}.
\end{aligned}
\tag{8.26}
$$

Materials for which the stress is derivable from a stored energy potential are called *hyperelastic* materials.

The governing equations are [recall (4.14)]

$$
\operatorname{Div}\left((\mathbf{I} + \nabla\boldsymbol{u})\frac{\partial W}{\partial \mathbf{E}}\bigg|_{\text{sym}}\right) + \mathbf{f}_0 = \varrho_0 \frac{\partial^2 \boldsymbol{u}}{\partial t^2},
\tag{8.27}
$$

with

$$
\begin{aligned}
\mathbf{F} &= (\mathbf{I} + \nabla\boldsymbol{u}), \\
\mathbf{S} &= \frac{\partial W}{\partial \mathbf{E}}\bigg|_{sym} = \mathcal{T}(I_{\mathbf{E}}, I\!\!I_{\mathbf{E}}, I\!\!I\!\!I_{\mathbf{E}}, \mathbf{E}), \\
\mathbf{E} &= \frac{1}{2}(\nabla\boldsymbol{u} + \nabla\boldsymbol{u}^T + \nabla\boldsymbol{u}^T\nabla\boldsymbol{u}).
\end{aligned}
\tag{8.28}
$$

Linear Elasticity   One of the most important and widely applied theories of continuum mechanics concerns the linearized version of the equations of elasticity arising from the assumptions of "small" strains. In this case,

$$
\begin{aligned}
\mathbf{E} &\approx e = \frac{1}{2}(\nabla\boldsymbol{u} + \nabla\boldsymbol{u}^T), \\
W &= \frac{1}{2}E_{ijk\ell}e_{k\ell}\,e_{ij}, \\
S_{ij} &= \frac{\partial W}{\partial e_{ij}} = E_{ijk\ell}e_{k\ell} = E_{ijk\ell}\frac{\partial u_k}{\partial X_\ell} \quad \text{(Hooke's Law)},
\end{aligned}
\tag{8.29}
$$

and we assume

$$
E_{ijk\ell} = E_{jik\ell} = E_{ij\ell k} = E_{k\ell ij}.
\tag{8.30}
$$

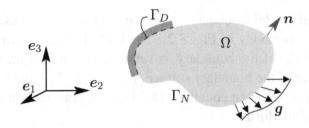

Figure 8.2: Elastic body in equilibrium.

Then, for small strains, we have

$$\frac{\partial}{\partial X_j}\left(E_{ijk\ell}\frac{\partial u_k}{\partial X_\ell}\right) + f_{0i} = \varrho_0\frac{\partial^2 u_i}{\partial t^2}. \tag{8.31}$$

For isotropic materials, we have

$$E_{ijk\ell} = \lambda\delta_{ij}\delta_{k\ell} + \mu(\delta_{ik}\delta_{j\ell} + \delta_{i\ell}\delta_{jk}), \tag{8.32}$$

where λ, μ are the Lamé constants:

$$\lambda = \frac{\nu E}{(1+\nu)(1-2\nu)}, \qquad \mu = \frac{E}{2(1+\nu)}, \tag{8.33}$$

E is Young's Modulus and ν Poisson's Ratio. Then

$$\mathbf{S} = \lambda\,(\mathrm{tr}\,\mathbf{E})\mathbf{I} + 2\mu\mathbf{E} = \lambda\,\mathrm{div}\,\boldsymbol{u}\,\mathbf{I} + 2\mu(\nabla\boldsymbol{u})_{sym}. \tag{8.34}$$

Suppose (8.34) holds and the body is in static equilibrium $(\partial^2\boldsymbol{u}/\partial t^2 = \mathbf{0})$. Then (8.31) reduces to the Lamé equations of elastostatics. In component form, these can be written

$$\lambda\frac{\partial^2 u_k}{\partial X_i\partial X_k} + \mu\left(\frac{\partial^2 u_i}{\partial X_k\partial X_k} + \frac{\partial^2 u_k}{\partial X_i\partial X_k}\right) = -f_{0i}, \quad 1 \le i, k \le 3. \tag{8.35}$$

A typical model boundary-value problem in linear elastostatics is suggested by the body in Fig. 8.2. The body is fixed against motions on a portion Γ_D of the boundary, is subjected to surface tractions g on the remainder of the boundary surface, $\Gamma_N = \partial\Omega \setminus \Gamma_D$, and is subjected to body forces f_0. Then equations (8.35) are augmented with boundary conditions of the form

$$
\begin{aligned}
u_i &= 0 \quad \text{on } \Gamma_D, \\
n_j E_{ijkl} \frac{\partial u_k}{\partial X_l} &= g_i \quad \text{on } \Gamma_N.
\end{aligned}
\tag{8.36}
$$

Part II

Electromagnetic Field Theory
and
Quantum Mechanics

CHAPTER 9

ELECTROMAGNETIC WAVES

9.1   Introduction

The now classical science of electricity and magnetism recognizes that material objects can possess what is called an *electric charge* — an intrinsic characteristic of the fundamental particles that make up the objects. The charges in objects we encounter in everyday life may not be apparent, because the object is *electrically neutral*, carrying equal amounts of two kinds of charges, *positive charge* and *negative charge*. Atoms consist of positively charged *protons*, negatively charged *electrons*, and electrically neutral *neutrons*, the protons and neutrons being packed together in the *nucleus* of the atom.

9.2   Electric Fields

The mathematical characterization of how charges interact with one another began with Coulomb's Law, postulated in 1783 by Charles Augustin de Coulomb on the basis of experiments (but actually discovered earlier by Henry Cavendish). It is stated as follows: Consider two charged particles (or point charges) of magnitude q_1 and q_2 separated by a distance r. The electrostatic force of attraction or repulsion between

An Introduction to Mathematical Modeling: A Course in Mechanics, First Edition. By J. Tinsley Oden
© 2011 John Wiley & Sons, Inc. Published 2011 by John Wiley & Sons, Inc.

the charges has magnitude

$$F = k \frac{|q_1| \, |q_2|}{r^2},$$

(9.1)

where k is a constant, normally expressed in the form

$$k = \frac{1}{4\pi\epsilon_0} = 8.99 \times 10^9 \text{ N m}^2/\text{C}^2,$$

where ϵ_0 is the permittivity constant,

$$\epsilon_0 = 8.85 \times 10^{-12} \left(\text{N m}^2/\text{C}^2\right)^{-1}.$$

Here C is the SI measure of a unit charge, called a Coulomb, and is defined as the amount of charge that is transferred across a material wire in one second due to a 1-ampere current in the wire. The reason for this choice of k is made clear later.

According to Jackson [29], the inverse-square dependence of force on distance (of Coulomb's Law) is known to hold over at least 24 orders of magnitude in the length scale.

Two important properties of charge are as follows:

1. Charge is *quantized*: Let e denote the elementary charge of a single electron or proton, known from experiments to be

$$e = 1.60 \times 10^{-19} \text{ C};$$

then any positive or negative charge q is of the form

$$q = \pm n e, \qquad n = 1, 2, 3, \dots \quad (n \in \mathbb{N}).$$

The fact that electric charge is "quantized" (meaning discretely defined as an integer multiple of e) is regarded by some as "one of the most profound mysteries of the physical world" (cf. Jackson [29, page 251]).

2. Charge is *conserved*: The net charge in a system or object is preserved, and it is constant unless additional charged particles are added to the system. We return to this idea in the next chapter.

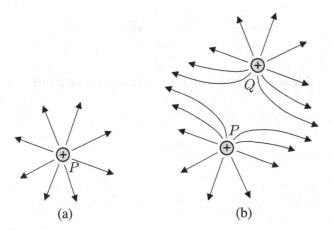

(a)   (b)

Figure 9.1: (a) A positive charge q at a point P in the plane, and electric field lines emanating from P. (b) The electric field produced by equal positive charges q_1 and q_2 at points P and Q.

The fundamental notion of an electric field is intimately tied to the force field generated by electric charges. Let q_1 denote a positive point charge situated at a point P in space. Imagine that a second positive point charge q_2 is placed at point Q near to P. According to Coulomb's Law, q_1 exerts a repulsive electrostatic force on q_2. This vector field of forces is called the *electric field*. We say that q_1 sets up an electric field $\boldsymbol{E}$ in the space surrounding it, such that the magnitude of $\boldsymbol{E}$ at a point Q depends upon the distance from P to Q and the direction depends on the direction from P to Q and the electrical sign of q_1 at P. In practice, $\boldsymbol{E}$ is determined at a point by evaluating the electrostatic force $\boldsymbol{F}$ due to a positive test charge q_0 at that point (see Fig. 9.1(a)). Then $\boldsymbol{E} = q_0^{-1}\boldsymbol{F}$. $\boldsymbol{E}$ thus has the units of Newtons per Coulomb (N/C). The magnitude of the electric field, then, due to a point charge q_1 is

$$|q_0|^{-1}\left(\frac{1}{4\pi\epsilon_0}\frac{|q_1|\,|q_0|}{r^2}\right) = \frac{|q_1|}{4\pi\epsilon_0\,r^2}.$$

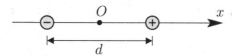

Figure 9.2: Equal and opposite charges on a line.

For m such charges,

$$E(\mathbf{x}) = q_0^{-1} \sum_{i=1}^{m} \mathbf{F}_{0i}(\mathbf{x}) = \frac{1}{4\pi\epsilon_0} \sum_{i=1}^{m} q_i \frac{\mathbf{x} - \mathbf{x}_i}{|\mathbf{x} - \mathbf{x}_i|^3}, \qquad (9.2)$$

$\mathbf{F}_{0i}$ being the forces from 0, the point of application of q_0, and the m charges q_i, $i = 1, 2, \ldots, m$, at points $\mathbf{x}_i$. The electric field lines for two positive charges are illustrated in Fig. 9.1(b).

A fundamental question arises concerning the electric force between two point charges q_1 and q_2 separated by a distance r; if q_2 is moved toward q_1, does the electric field change immediately? The answer is no. The information that one of the charges has moved travels outwardly in all directions at the speed of light c as an electromagnetic wave. More on this later.

The concept of an *electric dipole* is also important. The electric field of two particles of charge q but opposite sign, a distance d apart, at a point x on an axis through the point charges, is

$$E(x) = \frac{q}{4\pi\epsilon_0} \left(\frac{1}{(x - d/2)^2} - \frac{1}{(x + d/2)^2} \right) \boldsymbol{i},$$

$\boldsymbol{i}$ being a unit vector along the x-axis; see Fig. 9.2. For $x \gg d$, we have

$$E(x) = \frac{1}{2\pi\epsilon_0} \frac{qd}{x^3} \boldsymbol{i}.$$

Now let us examine the two-dimensional situation shown in Fig. 9.3 in which the two charged particles are immersed in an electric field $\boldsymbol{E}$ due to a remote charge. Clearly, a moment (a couple) is produced by the parallel oppositely directed forces of magnitude $|\mathbf{F}|$. This moment, denoted $\boldsymbol{m}$, is given by

$$\boldsymbol{m} = qd\,\boldsymbol{i} \qquad (9.3)$$

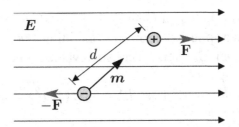

Figure 9.3: Equal and opposite charges in an electric field E a distance d apart.

and is called the *electric dipole moment*, and

$$E(x) = m \, / \, 2\pi\epsilon_0 \, x^3. \tag{9.4}$$

The torque τ created by a dipole moment in an electric field E is $m \times E$, as can be deduced from Fig. 9.3.

9.3 Gauss's Law

A fundamental property of charged bodies is that charge is *conserved*, as noted earlier. The charge of one body may be passed to another, but the net charge remains the same. No exceptions to this observation have ever been found. This idea of conservation of charge is embodied in *Gauss's Law*, which asserts that the net charge contained in a bounded domain is balanced by the flux of the electric field through the surface. This is merely a restatement of the property noted earlier that charge is conserved.

The product of the permeability and the net flux of an electric field E through the boundary surface $\partial\Omega$ of a bounded region Ω must be equal to the total charge q_Ω contained in Ω.

$$q_\Omega = \oint_{\partial\Omega} \epsilon_0 E \cdot n \, dA, \tag{9.5}$$

where n is a unit normal to $\partial\Omega$ and q_Ω is the net charge enclosed in the region Ω bounded by the surface $\partial\Omega$.

The relationship between Gauss's Law and Coulomb's Law is immediate: Let Ω be a sphere of radius r containing a positive point charge q at its center, and let $\boldsymbol{E}$ assume a constant value E on the surface. Then

$$\epsilon_0 \oint_{\partial(\text{sphere})} \boldsymbol{E} \cdot \boldsymbol{n} \, dA = \epsilon_0 E \left(4\pi r^2\right) = q,$$

so

$$E = \frac{1}{4\pi\epsilon_0} \frac{q}{r^2},$$

which is precisely Coulomb's Law. This relationship is the motivation for choosing the constant $k = 1/(4\pi\epsilon_0)$ in Coulomb's Law.

In many cases, the total charge q_Ω contained in region Ω is distributed so that it makes sense to introduce a *charge density* ρ representing the charge per unit volume. Then

$$q_\Omega = \int_\Omega \rho \, dx, \tag{9.6}$$

and Gauss's Law becomes

$$\oint_{\partial\Omega} \epsilon_0 \boldsymbol{E} \cdot \boldsymbol{n} \, dA = \int_\Omega \rho \, dx. \tag{9.7}$$

9.4   Electric Potential Energy

When an electrostatic force acts between two or more charged particles within a system of particles, an electrostatic potential energy U can be assigned to the system such that in any change ΔU in U the electrostatic forces do work. The potential energy per unit charge (or due to a charge q) is called the *electric potential* or voltage V; $V = U/q$. V has units of volts, 1 volt $= 1$ joule/coulomb.

9.4.1   Atom Models

The attractive or repulsive forces of charged particles lead directly to the classical *Rutherford model* of an atom as a tiny solar system in which

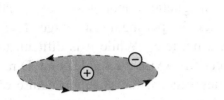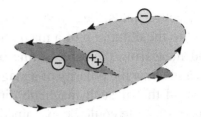

Figure 9.4: Model of an atom as charged electrons in orbits around a nucleus.

electrons, with negative charges e, move in orbits about a positively charged nucleus; see Fig. 9.4. Thus, when an electron travels close to a fixed positive charge $q = +e$, it can escape the pull or reach a stable orbit, spinning around the charge and thereby creating a primitive model of an atom. This primitive model was discarded when Bohr introduced the quantum model of atoms in which, for any atom, electrons can only exist in so-called discrete quantum states of well-defined energy or so-called energy shells. The motion of an electron from one shell to another can only happen instantaneously in a quantum jump. In such a jump, there is obviously a change in energy, which is emitted as a photon of electromagnetic radiation (more on this later). Bohr's model was later improved by the probability density model of Schrödinger, which we take up in the next chapter.

9.5 Magnetic Fields

Just as charged objects produce electric fields, magnets produce a vector field B called a *magnetic field*. Such fields are created by moving electrically charged particles or as an intrinsic property of elementary particles, such as electrons. In this latter case, magnetic fields are recognized as basic characteristics of particles, along with mass, electric charge, etc. In some materials the magnetic fields of all electrons cancel out, giving no net magnetic field. In other materials, they add together, yielding a magnetic field around the material.

There are no "monopole" analogies of magnetic fields as in the case of electric fields, i.e., there are no "magnetic monopoles" that would lead to the definition of a magnetic field by putting a test charge at rest and measuring the force acting on a particle. While it is difficult to imagine how the polarity of a magnet could exist at a single point, Dirac showed that if such magnetic dipoles existed, the quantum nature of electric charge could be explained. But despite numerous attempts, the existence of magnetic dipoles has never been conclusively proven. This fact has a strong influence on the mathematical structure of the theory of electromagnetism. Then, instead of a test charge, we consider a particle of charge q moving through a point P with velocity v. A force $\boldsymbol{F}_B$ is developed at P. The magnetic field $\boldsymbol{B}$ at P is defined as the vector field such that

$$\boldsymbol{F}_B = q\boldsymbol{v} \times \boldsymbol{B}. \tag{9.8}$$

Its units are *teslas*: T = newton/(coulomb)(meter/second) = N/(C/s)(m) or *gauss* (10^{-4} tesla). The force $\boldsymbol{F}$ is called the *Lorenz force*. In the case of a combined electric field $\boldsymbol{E}$ and magnetic field $\boldsymbol{B}$, we have

$$\boldsymbol{F}_B = q\big(\boldsymbol{E} + \boldsymbol{v} \times \boldsymbol{B}\big),$$

with v sometimes replaced by v/c, c being the speed of light in a vacuum. Since the motion of electric charges is called *current*, it is easily shown that $\boldsymbol{B}$ can be expressed in terms of the current i for various motions of charges. *Ampere's Law* asserts that on any closed loop $\mathcal{C}$ in a plane, we have

$$\oint_{\mathcal{C}} \boldsymbol{B} \cdot d\boldsymbol{s} = \mu_0 i_{\text{enclosed}},$$

where μ_0 is the *permeability constant*, $\mu_0 = 4\pi \times 10^{-7}$ T-m/A, and i_{enclosed} is the net current flowing perpendicular to the plane in the planar region enclosed by $\mathcal{C}$. Current is a phenomenon manifested by the flow of electric charges (charged particle), generally electrons, and, as a physical quantity, is a measure of the flow of electric charge. It is assigned the units A for ampere, a measure of the charge through a surface at a rate of one coulomb per second. For motion of a charge along a straight line, $|\boldsymbol{B}| = \mu_0 i / 2\pi R$, R being the perpendicular distance from the infinite line.

Ampere's Law does not take into account the induced magnetic field due to a change in the electric flux. When this is taken into account, the above equality is replaced by the *Ampere–Maxwell Law*,

$$\oint_{\mathcal{C}} \boldsymbol{B} \cdot d\boldsymbol{s} = \mu_0 i_{\text{enclosed}} + \mu_0 \epsilon_0 \frac{d}{dt} \Phi_E, \tag{9.9}$$

where Φ_E is the electric flux,

$$\Phi_E = \int_A \boldsymbol{E} \cdot \boldsymbol{n} \, dA,$$

$\boldsymbol{n}$ being a unit normal to the area A circumscribed by the closed loop $\mathcal{C}$. In keeping with the convention of defining a charge density ρ as in (9.6), we can also define a current flux density $\boldsymbol{j}$ such that

$$i_{\text{enclosed}} = \int_A \boldsymbol{j} \cdot \boldsymbol{n} \, dA. \tag{9.10}$$

Then (9.9) becomes

$$\oint_{\mathcal{C}} \boldsymbol{B} \cdot d\boldsymbol{s} = \mu_0 \int_A \boldsymbol{j} \cdot \boldsymbol{n} \, dA + \mu_0 \epsilon_0 \frac{d}{dt} \int_A \boldsymbol{E} \cdot \boldsymbol{n} \, dA.$$

Just as the motion of a charged particle produces a magnetic field, so also does the change of a magnetic field produce an electric field. This is called an *induced electric field* and is characterized by *Faraday's Law*: Consider a particle of charge q moving around a closed loop $\mathcal{C}$ encompassing an area A with unit normal $\boldsymbol{n}$. Then $\boldsymbol{B}$ induces an electric field $\boldsymbol{E}$ such that

$$\oint_{\mathcal{C}} \boldsymbol{E} \cdot d\boldsymbol{s} = -\frac{d}{dt} \int_A \boldsymbol{B} \cdot \boldsymbol{n} \, dA. \tag{9.11}$$

From the fact that magnetic materials have poles of attraction and repulsion, it can be appreciated that magnetic structure can exist in the form of magnetic dipoles. Magnetic monopoles do not exist. For this

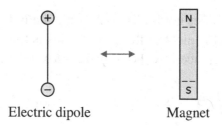

Electric dipole   Magnet

Figure 9.5: Electric dipole analogous to a dipole caused by a magnet.

reason, the net magnetic flux through a closed Gaussian surface must be zero:

$$\oint_{\partial\Omega} \boldsymbol{B} \cdot \boldsymbol{n} \, dA = 0. \tag{9.12}$$

This is referred to as Gauss's Law for magnetic fields.

Every electron has an intrinsic angular momentum $\boldsymbol{s}$, called the *spin angular momentum*, and an intrinsic *spin magnetic dipole moment* $\boldsymbol{\mu}$, related by

$$\boldsymbol{\mu} = -\frac{e}{m}\boldsymbol{s}, \tag{9.13}$$

where e is the elementary charge (1.6×10^{-19} C) and m is the mass of an electron (9.11×10^{-31} kg).

9.6 Some Properties of Waves

The concept of a wave is a familiar one from everyday experiences. In general, a wave is a perturbation or disturbance in some physical quantity that propagates in space over some period of time. Thus, an acoustic wave represents the space and time variation of pressure perturbations responsible for sound; water waves represent the motion of the surface of a body of water over a time period; electromagnetic waves, as will be established, characterize the evolution of electric and magnetic fields over time, but need no media through which to move as they propagate in a perfect vacuum at the speed of light.

Mathematically, we can characterize a wave by simply introducing a function u of position x (or $\mathbf{x} = (x_1, x_2, x_3)$ in three dimensions) and time t. In general, waves can be represented as the superposition of simple sinusoidal functions, so that the building blocks for wave theory are functions of the form

$$u(x, t) = u_0 e^{i(kx - \omega t)}, \qquad (9.14)$$

or, for simplicity, of the form

$$u(x, t) = u_0 \sin(kx - \omega t), \qquad (9.15)$$

which are called plane waves. A plot of this last equation is given in Fig. 9.7. Here

$$u_0 = \text{the amplitude of the wave,}$$
$$k = \text{the angular wave number,} \qquad (9.16)$$
$$\omega = \text{the angular frequency.}$$

We also define

$$\lambda = 2\pi/k = \text{the wavelength,}$$
$$T = 2\pi/\omega = \text{the period (of oscillation).} \qquad (9.17)$$

The frequency ν of the wave is defined as

$$\nu = 1/T. \qquad (9.18)$$

The *wave speed* v is defined as the rate at which the wave pattern moves, as indicated in Fig. 9.6. Since point A retains its position on the wave crest as the wave moves from left to right, the quantity

$$\varphi = kx - \omega t$$

must be constant. Thus

$$\frac{d\varphi}{dt} = 0 = k\frac{dx}{dt} - \omega,$$

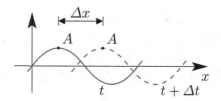

Figure 9.6: Incremental motion of a wave front over a time increment Δt.

so the wave speed is

$$v = \frac{\omega}{k} = \frac{\lambda}{T} = \lambda\nu. \tag{9.19}$$

The quantity

$$\varphi = kx - \omega t \tag{9.20}$$

is the phase of the wave. Two waves of the form

$$u_1 = u_0 \sin(kx - \omega t) \qquad \text{and} \qquad u_2 = u_0 \sin(kx - \omega t + \varphi)$$

have the same amplitude, angular frequency, angular wave number, wavelength, and period, but are *out of phase* by φ. These waves produce *interference* when superimposed:

$$\begin{aligned}
u(x,t) &= u_1(x,t) + u_2(x,t) \\
&= (2u_0 \cos \varphi/2) \sin(kx - \omega t + \varphi/2). \tag{9.21}
\end{aligned}$$

For

- $\varphi = 0$, the amplitude doubles while the wavelength and period remains the same (constructive interference);

- $\varphi = \pi$, the waves cancel ($u(x,t) \equiv 0$) (destructive interference).

Observe that

$$\frac{\partial^2 u}{\partial t^2} = -\omega^2 u_0 \sin(kx - \omega t)$$

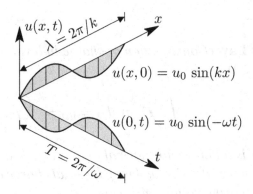

Figure 9.7: Properties of a simple plane wave of the form $u(x,t) = u_0 \sin(kx - \omega t)$.

and

$$\frac{\partial^2 u}{\partial x^2} = -k^2 u_0 \sin(kx - \omega t),$$

so that u satisfies the second-order (hyperbolic) wave equation

$$\frac{\partial^2 u}{\partial t^2} - \left(\frac{\omega^2}{k^2}\right)\frac{\partial^2 u}{\partial x^2} = 0, \tag{9.22}$$

and, again, ω/k is recognized as the wave speed.

9.7 Maxwell's Equations

Consider electric and magnetic fields E and B in a vacuum. Recall the following physical laws:

Gauss's Law *(Conservation of charge in a volume Ω enclosed by a surface $\partial\Omega$)*

$$\epsilon_0 \int_{\partial\Omega} \boldsymbol{E} \cdot \boldsymbol{n} \, dA = q_\Omega = \int_\Omega \rho \, dx, \qquad (9.23)$$

where $\boldsymbol{n}$ is a unit vector normal to the surface area element dA and ρ is the charge density, q_Ω being the total charge contained in Ω, and $dx = d^3x$ is the volume element.

Faraday's Law *(The induced electromotive force due to a charge in magnetic flux)*

$$\int_{\mathcal{C}} \boldsymbol{E} \cdot d\boldsymbol{s} = -\frac{d}{dt} \int_A \boldsymbol{B} \cdot \boldsymbol{n} \, dA, \qquad (9.24)$$

where $\mathcal{C}$ is a closed loop surrounding a surface of area A and $\boldsymbol{n}$ is a unit normal to dA. The total electromotive force is $\int_{\mathcal{C}} \boldsymbol{E} \cdot d\boldsymbol{s}$.

The Ampere–Maxwell Law *(The magnetic field produced by a current i)*

$$\int_{\mathcal{C}} \boldsymbol{B} \cdot d\boldsymbol{s} = \mu_0 i_{\text{enclosed}} + \mu_0 \epsilon_0 \frac{d}{dt} \left(\int_A \boldsymbol{E} \cdot \boldsymbol{n} \, dA \right)$$

$$= \mu_0 \int_A \boldsymbol{j} \cdot \boldsymbol{n} \, dA + \mu_0 \epsilon_0 \int_A \boldsymbol{n} \cdot \frac{\partial \boldsymbol{E}}{\partial t} \, dA, \qquad (9.25)$$

where $\mathcal{C}$ is a closed curve surrounding a surface of area A, $\boldsymbol{j}$ is the current density, and $\boldsymbol{n}$ is a unit vector normal to dA.

The Absence of Magnetic Monopoles

$$\int_{\partial\Omega} \boldsymbol{B} \cdot \boldsymbol{n}\, dA = 0, \tag{9.26}$$

where $\partial\Omega$ is a surface bounding a bounded region $\Omega \subset \mathbb{R}^3$.

Applying the divergence theorem to the left-hand sides of (9.23) and (9.26), applying Stokes' theorem to the left-hand sides of (9.24) and (9.25), and arguing that the integrands of the resulting integrals must agree almost everywhere, one gets the following system of equations:

$$\epsilon_0 \boldsymbol{\nabla} \cdot \boldsymbol{E} = \rho,$$

$$\boldsymbol{\nabla} \times \boldsymbol{E} = -\frac{\partial \boldsymbol{B}}{\partial t},$$

$$\boldsymbol{\nabla} \times \boldsymbol{B} = \mu_0 \boldsymbol{j} + \mu_0 \epsilon_0 \frac{\partial \boldsymbol{E}}{\partial t},$$

$$\boldsymbol{\nabla} \cdot \boldsymbol{B} = 0. \tag{9.27}$$

These are Maxwell's equations of classical electromagnetics of macroscopic events in a vacuum. Here we do not make a distinction between ∇ and grad, Div and div, etc.

In materials other than a vacuum, the permittivity ϵ and the magnetic permeability μ may be functions of position x in Ω and must satisfy constitutive equations

$$\begin{aligned} \boldsymbol{D} &= \epsilon_0 \boldsymbol{E} + \boldsymbol{P} &&= \text{the electric displacement field,}\\ \boldsymbol{H} &= \mu_0^{-1} \boldsymbol{B} - \boldsymbol{M} &&= \text{the magnetic field,} \end{aligned} \tag{9.28}$$

with $\boldsymbol{B}$ now called the magnetic inductance, $\boldsymbol{P}$ the polarization vector and $\boldsymbol{M}$ the magnetic dipole. Then the equations are rewritten in terms of D, H, B, and j:

$$\nabla \cdot D = \rho_f,$$

$$\nabla \times E = -\frac{\partial B}{\partial t},$$

$$\nabla \cdot B = 0,$$

$$\nabla \times H = j_f + \frac{\partial D}{\partial t},$$

(9.29)

where ρ_f and j_f are appropriately scaled charge and current densities. When quantum and relativistic effects are taken into account, an additional term appears in the first and third equations and $\partial/\partial t$ is replaced by the total material time derivative,

$$\frac{d}{dt} = \frac{\partial}{\partial t} + v \cdot \nabla,$$

v being the velocity of the media.

Let $j = 0$. Then

$$\frac{\partial}{\partial t}\nabla \times B = \nabla \times \frac{\partial B}{\partial t} = -\nabla \times \nabla \times E = \mu_0\epsilon_0\frac{\partial^2 E}{\partial t^2},$$

so that we arrive at the wave equation

$$\frac{\partial^2 E}{\partial t^2} + \frac{1}{\mu_0\epsilon_0}\nabla \times \nabla \times E = 0.$$

(9.30)

From the fact that $\nabla \times \nabla \times E = \nabla(\nabla \cdot E) - \Delta E = \nabla(\rho/\epsilon_0) - \Delta E$, with Δ being the Laplacian, we arrive at the wave equation

$$\frac{\partial^2 E}{\partial t^2} - \frac{1}{\mu_0\epsilon_0}\Delta E = -\frac{1}{\epsilon_0}\nabla\rho.$$

(9.31)

Comparing this with (9.22), we see that the speed of propagation of electromagnetic disturbances is precisely

$$c = \frac{1}{\sqrt{\mu_0\epsilon_0}} = 3.0 \times 10^8 \text{ m/s},$$

(9.32)

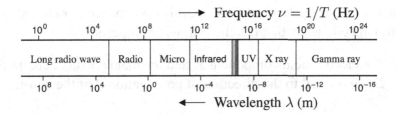

Figure 9.8: The electromagnetic wave spectrum.

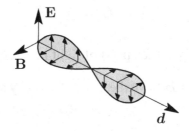

Figure 9.9: Components of an electromagnetic plane wave propagating in the direction d.

which is the speed of light in a vacuum, i.e.,

> *Electromagnetic disturbances (electromagnetic waves) travel (radiate) at the speed of light through a vacuum.*

9.8 Electromagnetic Waves

Consider a disturbance of an electric (or magnetic) field. Of various wave forms, we consider here plane waves, which are constant frequency waves whose fronts (surfaces of constant phase) are parallel planes of constant amplitude normal to the direction of propagation. According to what we have established thus far, this disturbance is radiated as an electromagnetic wave that has the following properties:

- The electric field component E and the magnetic field component B are normal to the direction of propagation d of the wave:

$$d = E \times B / |E \times B|.$$

- E is normal to B:

$$E \cdot B = 0.$$

- The wave travels at the speed of light in a vacuum.

- The fields E and B have the same frequency and are in phase with one another.

Thus, while the wave speed $c = \omega/k = \lambda/T$ is constant, the wavelength λ can vary enormously. The well-known scales of electromagnetic wavelengths is given in Fig. 9.8.

Some electromagnetic waves such as X rays and visible light are radiated from sources that are of atomic or nuclear dimension. The quantum transfer of an electron from one shell to another radiates an electromagnetic wave. The wave propagates an energy packet called a *photon* in the direction d of propagation. We explore this phenomenon in more detail in the next chapters.

CHAPTER 10

INTRODUCTION TO QUANTUM MECHANICS

10.1 Introductory Comments

Quantum mechanics emerged as a theory put forth to resolve two fundamental paradoxes that arose in describing physical phenomena using classical physics. First, can physical events be described by waves (optics, wave mechanics) or particles (the mechanics of "corpuscles" or particles) or both? Second, at atomic scales, experimental evidence confirms that physical quantities may take on only discrete values; they are said to be *quantized*, not continuous in time, and thus can take on instantaneous jumps in values. Waves are characterized by frequency ν and wave number k (or period $T = 1/\nu$ and wavelength $\lambda = 2\pi/k$) while the motion of a particle is characterized by its total energy E and its momentum p. The resolution of this wave-particle paradox took place through an amazing sequence of discoveries that began at the beginning of the twentieth century and led to the birth of quantum mechanics, with the emergence of Schrödinger's equation in 1926 and additional refinements by Dirac and others in subsequent years. Our goal in this chapter is to present the arguments and derivations leading to Schrödinger's equation and to expose the dual nature of wave and particle descriptions

An Introduction to Mathematical Modeling: A Course in Mechanics, First Edition. By J. Tinsley Oden
© 2011 John Wiley & Sons, Inc. Published 2011 by John Wiley & Sons, Inc.

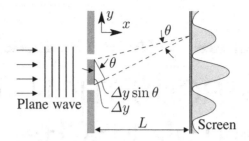

Figure 10.1: Diffraction of a beam of light.

of physical events.

10.2 Wave and Particle Mechanics

We have seen in the previous chapter that basic physical events observed in nature can be explained using the concept of waves—from optics and the understanding of light to radio waves, X rays, and so on. Waves propagating at the same phase have the superposition property described earlier and, because of this property, explain what is known as light diffraction, illustrated in Fig. 10.1. There we see depicted the so-called double-slit experiment, in which a beam of light is produced which is incident on two thin slits in a wall a distance Δy apart. The light, which we know to be an electromagnetic wave at a certain wavelength λ, leaves the slit at a diffraction angle θ and strikes a screen parallel to the wall making a pattern of dark and light fringes on the screen as waves of different phases reinforce or cancel one another. The difference in path lengths of waves leaving the upper and lower slits is $\delta = \Delta y \sin \theta = \xi \lambda$, where ξ is an integer n if it corresponds to a light fringe and $\xi = n + \frac{1}{2}$ if it corresponds to a dark fringe. Furthermore, the distance on the screen between two light fringes is $\lambda L / \Delta y$, L being the distance from the slits to the screen. These types of observations can be regarded as consistent and predictable by wave theory. We return to this observational setup later.

The classical theory of electromagnetic waves leading to Maxwell's equations described in the previous chapter, which is firmly based on the notion of waves, became in dramatic conflict with physical experiments around the end of the nineteenth century. Among observed physical phenomena that could not be explained by classical wave theory was the problem of blackbody radiation, which has to do with the radiation of light emitted from a heated solid. Experiments showed that classical electromagnetic theory could not explain observations at high frequencies. In 1900, Planck introduced a theory that employed a dramatic connection between particles and waves and correctly explained the experimental observations. He put forth the idea that exchanges in energy E (a particle theory concept) do not occur in a continuous manner predicted by classical theory, but are manifested in discrete packets of magnitude proportional to the frequency:

$$\boxed{E = h\nu.} \tag{10.1}$$

The constant of proportionality h is known as Planck's constant. Since $\nu = \omega/2\pi$, we have $E = \hbar\omega$ where

$$\hbar = \frac{h}{2\pi} = 1.05457 \times 10^{-34}\,\text{J s}.$$

The constant h was determined by adjusting it to agree with experimental data on the energy spectrum.

According to Messiah [37, p. 11], the general attitude toward Planck's result was that "everything behaves as if" the energy exchange between radiation and the black box occur by quanta. Half a decade later (in 1905), Einstein's explanation of the photoelectric effect confirmed and generalized Planck's theory and established the fundamental quantum structure of the energy–light–wave relationship. The photoelectric effect is observed in an experiment in which an alkali metal is subjected to short wavelength light in a vacuum. The surface becomes positively charged as it gives off negative electricity in the form of electrons. Very precise experiments enable one to measure the total current and the velocity of the electrons. Observations confirm that:

1. the velocity of the emitted electrons depends only on the frequency ν of the light (the electromagnetic wave);

2. the emission of electrons is observed immediately at the start of radiation (contrary to classical theory);

3. the energy of the electron is, within a constant $h\nu_0$ depending on the material, proportional to the frequency ν, with constant of proportionality h, Planck's constant:

$$E = h\nu - h\nu_0.$$

This last result is interpreted as follows: Every light quantum striking the metal collides with an electron to which it transfers its energy. The electron loses part of its energy $(-h\nu_0)$ due to the work required to remove it from the metal. This is the so-called binding energy of the material defined by a threshold frequency ν_0 which must be exceeded before the emission of the electrons takes place. The velocity of the electron does not depend on the intensity of the light, but the number of electrons emitted does, and is equal to the number of incident light quanta.

Property 3 provides a generalization of Planck's theory which Einstein used to explain the photoelectric effect. Experiments by Meyer and Gerlach in 1914 were in excellent agreement with this proposition. The conclusion can be summarized as follows:

A wave of light, which is precisely an electromagnetic wave of the type discussed in Chapter 9, can be viewed as a stream of discrete energy packets called photons, *which can impart to electrons an instantaneous quantum change in energy of magnitude* $h\nu$.

This fact led to significant revisions in the model of atomic structure.

Just as the theories of Planck and Einstein revealed the particle nature of waves, the wave nature of particles was laid down in the dissertation of de Broglie in 1923. There it was postulated that electrons and other particles have waves associated with them of wavelength λ proportional to the reciprocal of the particle momentum p, the constant of proportionality again being Planck's constant:

$$\lambda = \frac{h}{p}. \tag{10.2}$$

The de Broglie theory suggests that the diffraction of particles must be possible in the same spirit as wave diffraction, a phenomenon that was eventually observed by Davisson and Germer in 1927.

We observe that since photons travel at the speed of light c in a vacuum, the momentum p of a photon and its energy E are related by $E = pc = p\lambda\nu$, so $p = h/\lambda$ in agreement with the de Broglie relation (10.2). By a similar argument employing relativity theory, Einstein showed that photons have zero mass, a fact also consistent with experiments. Experimental evidence on frequencies or equivalently energy spectra of atoms also confirms the existence of discrete spectra, in conflict with the classical Rutherford model which assumes a continuous spectrum. Thus, discontinuities are also encountered in the interaction of particles (matter) with light.

10.3   Heisenberg's Uncertainty Principle

The Planck–Einstein relation (10.1) and the de Broglie relation (10.2) establish that physical phenomena can be interpreted in terms of particle properties or wave properties. Relation (10.2), however, soon led to a remarkable observation. In 1925, Heisenberg realized that it is impossible to determine both the position and momentum of a particle, particularly an electron, simultaneously. Wave and particle views of natural events are said to be in *duality* (or to be complementary) in that if the particle character of an experiment is proved, it is impossible to prove at the same time its wave character. Conversely, proof of a wave characteristic means that the particle characteristic, at the same time, cannot be established.

A classical example illustrating these ideas involves the diffraction of a beam of electrons through a slit, as indicated in Fig. 10.1. As the electrons pass through the slit, of width Δy, the beam is accompanied by diffraction and diverges by an angle θ. The electron is represented by a de Broglie wave of wavelength $\lambda = h/p$. Thus,

$$\Delta y \sin \theta \approx \lambda = h/p.$$

Likewise, the change in momentum in the y direction is

$$\Delta p = \Delta p_y = p \sin \theta.$$

Thus,

$$\boxed{\Delta y \, \Delta p_y \approx h.} \tag{10.3}$$

The precise position of an electron in the slit cannot thus be determined, nor can the variation in the momentum be determined with greater precision than Δp_y (or $h/\Delta y$).

This relation is an example of Heisenberg's uncertainty principle. Various other results of a similar structure can be deduced from quantum mechanics. Some examples and remarks follow:

1. A more precise analysis yields

$$\Delta y = \sigma_y = \sqrt{\langle y^2 \rangle - \langle y \rangle^2}, \qquad \Delta p = \sigma_p = \sqrt{\langle p^2 \rangle - \langle p \rangle^2},$$

$$\Delta y \, \Delta p = \sigma_y \sigma_p \geq \frac{1}{2}\hbar, \tag{10.4}$$

   where $\langle y \rangle$ is the average measurement of position around y and similar definitions apply to p, and σ_y and σ_p are the corresponding standard deviations in y and p, respectively.

2. Also, it can be shown that

$$\Delta t \, \Delta E \geq \frac{1}{2}\hbar. \tag{10.5}$$

   Thus, just as position and momentum cannot be localized in time, the time interval over which a change in energy occurs and the corresponding change in energy cannot be determined simultaneously.

We return to this subject and provide a more general version of the principle in Chapter 11.

10.4 Schrödinger's Equation

At this point, we are provided with the fundamental wave-particle relationships provided by the Planck–Einstein relation (10.1), the de Broglie relation (10.2), and the Heisenberg uncertainty principle (e.g., (10.4) or (10.5)). These facts suggest that to describe completely the behavior of a physical system, particularly an electron in motion, one could expect that a wave function of some sort should exist, because the dynamics of a particle is known to be related to properties of a wave. The location of the particle is uncertain, so the best that one could hope for is to know the probability that a particle is at a place x in space at a given time t. So the wave function should determine in some sense such a probability density function. Finally, the wave parameters defining the phase of the wave (e.g., k, ω or λ, ν) must be related to those of a particle (e.g., E, p) by relations (10.1) and (10.2).

10.4.1 The Case of a Free Particle

Considering first the wave equation of a free particle (ignoring hereafter relativistic effects), we begin with the fact that a wave $\Psi = \Psi(x, t)$ is a superposition of monochromatic plane waves (here in one space dimension). As we have seen earlier, plane waves are of the form

$$\Psi(x, t) = \psi_0 e^{i(kx - \omega t)},$$

where ψ_0 is the amplitude. Then a general wave function will be a superposition of waves of this form. Since now the wave number k and the angular frequency ω are related to energy and momentum according to $k = 2\pi/\lambda = p/\hbar$ and $\omega = 2\pi\nu = E/\hbar$, we have

$$\Psi(x, t) = \psi_0 e^{i(px - Et)/\hbar}.$$

Thus,

$$\frac{\partial \Psi}{\partial t} = -\frac{i}{\hbar} E\, \psi_0 e^{i(px - Et)/\hbar} = -\frac{i}{\hbar} E\, \Psi,$$

$$\frac{\partial \Psi}{\partial x} = \frac{i}{\hbar} p\, \psi_0 e^{i(px - Et)/\hbar} = \frac{i}{\hbar} p\, \Psi.$$

Thus, E and p can be viewed as operators on Ψ:

$$
\begin{aligned}
E\Psi &= \left(-\frac{\hbar}{i}\frac{\partial}{\partial t}\right)\Psi, \\
p\Psi &= \left(\frac{\hbar}{i}\frac{\partial}{\partial x}\right)\Psi.
\end{aligned}
\tag{10.6}
$$

For a free particle of mass m, the total energy E is kinetic energy and it is related to the momentum p by

$$
E = \frac{p^2}{2m}.
\tag{10.7}
$$

Thus, introducing (10.6) into (10.7), we arrive at the following partial differential equation for the wave functions:

$$
i\hbar\frac{\partial\Psi}{\partial t} + \frac{\hbar^2}{2m}\frac{\partial^2\Psi}{\partial x^2} = 0.
\tag{10.8}
$$

This is *Schrödinger's equation* for a free particle.

Of course, this notion of superposition of plane waves assumes that the support of Ψ is bounded, while we know this is not the case. So we may extend the domain to $\mathbb{R} = (-\infty, +\infty)$ and obtain the Fourier transform-type superposition of monochromatic plane waves,

$$
\Psi(x, t) = \int_{-\infty}^{\infty} \varphi(p)e^{i(px - Et)/\hbar}\, dp,
\tag{10.9}
$$

from which we obtain

$$
\begin{aligned}
i\hbar\frac{\partial}{\partial t}\Psi(x, t) &= \int_{-\infty}^{\infty} E\,\varphi(p)e^{i(px - Et)/\hbar}\, dp, \\
-\hbar^2\frac{\partial^2}{\partial x^2}\Psi(x, t) &= \int_{-\infty}^{\infty} p^2\,\varphi(p)e^{i(px - Et)/\hbar}\, dp.
\end{aligned}
\tag{10.10}
$$

The relation $E = p^2/2m$ then leads again to (10.8) for $x \in \mathbb{R}$. We interpret $\varphi(p)$ subsequently.

The wave function *completely determines the dynamical state of the quantum system* in the sense that all information on the state of the system can be deduced from the knowledge of Ψ. The central goal of quantum theory is this: Knowing Ψ at an initial time t_0, determine Ψ at later times t. To accomplish this, we must solve Schrödinger's equation.

10.4.2 Superposition in $\mathbb{R}^n$

Let us now consider the n-dimensional case, $n = 1, 2$ or 3. Now px becomes $\mathbf{p} \cdot \mathbf{x}$, $\mathbf{p} = (p_1, p_2, p_3)$ being the momentum vector. We can write Ψ in the form

$$\Psi(\mathbf{x}, t) = \psi_0 e^{i(\mathbf{p} \cdot \mathbf{x} - Et)/\hbar},$$

and obtain Schrödinger's equation,

$$\boxed{i\hbar \frac{\partial}{\partial t} \Psi(\mathbf{x}, t) + \frac{\hbar^2}{2m} \Delta \Psi(\mathbf{x}, t) = 0,} \qquad (10.11)$$

where Δ is the Laplacian

$$\Delta = \frac{\partial^2}{\partial x_1^2} + \frac{\partial^2}{\partial x_2^2} + \frac{\partial^2}{\partial x_3^2}.$$

We can express Ψ as a superposition of an infinite number of waves of amplitude φ according to

$$\Psi(\mathbf{x}, t) = \int_{\mathbb{R}^n} \varphi(\mathbf{p}) e^{i(\mathbf{p} \cdot \mathbf{x} - Et)/\hbar} d\mathbf{p},$$

with $d\mathbf{p} = dp_1 dp_2 dp_3$. What is φ? Let $t = 0$. Then the initial condition is

$$\Psi(\mathbf{x}, 0) = \varphi(\mathbf{x}) = \int_{\mathbb{R}^n} \varphi(\mathbf{p}) e^{i\mathbf{p} \cdot \mathbf{x}} d\mathbf{p},$$

i.e., $\varphi(\mathbf{p})$ is the Fourier transform (to within an irrelevant multiplication constant) of the initial data $\Psi(\mathbf{x}, 0)$.

10.4.3 Hamiltonian Form

A popular way of writing Schrödinger's equation is to introduce the Hamiltonian,

$$H(q,p)\Psi = E\Psi \qquad (q \sim x).$$

Then the Hamiltonian operator is written

$$H(q,p) = H\left(q, \frac{\hbar}{2\pi i}\frac{\partial}{\partial q}\right),$$

and Schrödinger's equation becomes

$$\left(H\left(q, \frac{\hbar}{i}\frac{\partial}{\partial q}\right) + \frac{\hbar}{i}\frac{\partial}{\partial t}\right)\Psi = 0. \qquad (10.12)$$

10.4.4 The Case of Potential Energy

If the particle is under the action of a force with potential energy $V = V(q)$, then

$$H(q,p) = \frac{p^2}{2m} + V(q), \qquad (10.13)$$

with q the coordinate of the particle, and Schrödinger's equation becomes

$$\left(-\frac{\hbar^2}{2m}\frac{\partial^2}{\partial q^2} + V(q) - i\hbar\frac{\partial}{\partial t}\right)\Psi = 0. \qquad (10.14)$$

For a particle moving in $\mathbb{R}^3$, this becomes

$$\left(-\frac{\hbar^2}{2m}\Delta + V(\boldsymbol{q}) - i\hbar\frac{\partial}{\partial t}\right)\Psi = 0. \qquad (10.15)$$

10.4.5 Relativistic Quantum Mechanics

While we are not going to cover relativistic effects, the wave equation for this case easily follows from the fact that in this case, the energy of a free particle is given by

$$E^2 = p^2c^2 + m^2c^4.$$

From this we deduce the equation

$$\left(\frac{1}{c^2}\frac{\partial^2}{\partial t^2} - \Delta + \left(\frac{mc}{\hbar}\right)^2\right)\Psi(\mathbf{x}, t) = 0. \tag{10.16}$$

This is called the Klein–Gordon equation. The Klein–Gordon extension does not take into account the intrinsic spin of elementary particles such as an electron. Dirac introduced spin as an additional degree of freedom that allowed the square root of E^2 to be computed, resulting in the relativistic quantum mechanics equation called the Dirac equation. In this setting, the wave function is a four-component spinor rather than a scalar-valued function. In nonrelativistic quantum mechanics, to which we restrict ourselves, the property of particle spin is introduced as a postulate. We discuss this subject in more detail in Chapter 13.

10.4.6 General Formulations of Schrödinger's Equation

The multiple particle case can be constructed in analogy with the one-dimensional free particle case. The basic plan is to consider a dynamical system of N particles with coordinates $q_1, q_2, \ldots, q_N$ and momenta $\mathbf{p}_1, \mathbf{p}_2, \ldots, \mathbf{p}_N$ for which the Hamiltonian is a functional,

$$H = H(q_1, q_2, \ldots, q_N; \mathbf{p}_1, \mathbf{p}_2, \ldots, \mathbf{p}_N; t).$$

To this dynamical system there corresponds a quantum system represented by a wave function $\Psi(q_1, q_2, \ldots, q_N, t)$. Setting

$$E = i\hbar\frac{\partial}{\partial t} \qquad \text{and} \qquad \mathbf{p}_r = \frac{\hbar}{i}\frac{\partial}{\partial q_r}; \qquad r = 1, 2, \ldots, N, \tag{10.17}$$

Schrödinger's equation becomes

$$\left(i\hbar\frac{\partial}{\partial t} - H\left(q_1, q_2, \ldots, q_N; \frac{\hbar}{i}\frac{\partial}{\partial q_1}, \frac{\hbar}{i}\frac{\partial}{\partial q_2}, \ldots, \frac{\hbar}{i}\frac{\partial}{\partial q_N}\right)\right)$$
$$\Psi(q_1, q_2, \ldots, q_N, t) = 0. \tag{10.18}$$

The rules (10.17) showing the correspondence of energy and momentum to the differential operators hold only in Cartesian coordinates in the form indicated, as Schrödinger's equation should be invariant under a rotation of the coordinate axes.

10.4.7   The Time-Independent Schrödinger Equation

In general, the wave function can be written as the product of a spatial wave function $\psi(\mathbf{x})$ and a harmonic time-dependent component,

$$\Psi(\mathbf{x}, t) = \psi(\mathbf{x})e^{-iEt/\hbar}. \tag{10.19}$$

That this product form is true in general can be verified by assuming that Ψ is the product of $\psi(\mathbf{x})$ and an arbitrary function $\varphi(t)$. Introducing $\psi\varphi$ into (10.15), dividing the resulting equation by $\psi\varphi$, and making the classical argument that a function of $\mathbf{x}$ can equal a function of t only if both are equal to a constant, we conclude that $\varphi(t)$ must be of the form $\varphi(t) = \exp(iEt/\hbar)$, hence (10.19).

From (10.19), we have

$$i\hbar\frac{\partial\Psi}{\partial t} = -i\hbar\,\frac{i}{\hbar}E\Psi = E\Psi.$$

Hence,

$$\boxed{H\psi = E\psi.} \tag{10.20}$$

This is the time-independent Schrödinger equation. It establishes that the energy E is an eigenvalue of the Hamiltonian operator and that the spatial wave function ψ is an eigenfunction of the Hamiltonian.

10.5   Elementary Properties of the Wave Equation

We will undertake a basic and introductory study of Schrödinger's equation, first for a single particle in one space dimension, and then generalize the analysis by considering a more general mathematical formalism provided by function space settings. An excellent source for this level of treatment is Griffiths's book, *Introduction to Quantum Mechanics* [21], but the books of Born [11] and Messiah [37] may also be consulted.

10.5.1   Review

The Schrödinger equation governing the dynamics of a single particle of mass m moving along a line, the x axis, subjected to forces derived from a potential V is

$$ i\hbar \frac{\partial \Psi}{\partial t} + \frac{\hbar^2}{2m} \frac{\partial^2 \Psi}{\partial x^2} - V\Psi = 0, \qquad (10.21) $$

where

$$ \hbar = \frac{h}{2\pi} = \text{Planck's constant}/2\pi = 1.054573 \times 10^{-34}\,\text{J s}, $$
$$ \Psi = \Psi(x, t) = \text{the wave function}. $$

To solve (10.21), we must add an initial condition,

$$ \Psi(x, 0) = \varphi(x), \qquad (10.22) $$

where φ is prescribed. We postulate that the wave function has the property

$$ \Psi^*\Psi = |\Psi(x, t)|^2 = \rho(x, t), $$

where Ψ^* is the complex conjugate of Ψ and

$$ \rho(x, t) = \text{the probability distribution function} $$
$$ \text{associated with } \Psi, \text{ such that} $$
$$ \rho(x, t)\,dx = \text{the probability of finding the particle} $$
$$ \text{between } x \text{ and } x + dx \text{ at time } t. $$

The wave function must be normalized since ρ is a PDF (a probability density function):

$$ \int_{-\infty}^{\infty} |\Psi(x, t)|^2 \, dx = 1. \qquad (10.23) $$

Proposition 10.1 *If* $\Psi = \Psi(x, t)$ *is a solution of Schrödinger's equation (10.21), then*

$$ \frac{d}{dt} \int_{-\infty}^{\infty} |\Psi(x, t)|^2 \, dx = 0. \qquad (10.24) $$

Proof

$$\frac{d}{dt} \int_{-\infty}^{\infty} |\Psi|^2 \, dx = \int_{-\infty}^{\infty} \frac{\partial}{\partial t} \left(\Psi^*(x,t)\Psi(x,t) \right) dx$$

$$= \int_{-\infty}^{\infty} \left(\Psi^* \frac{\partial \Psi}{\partial t} + \frac{\partial \Psi^*}{\partial t} \Psi \right) dx.$$

But

$$\frac{\partial \Psi}{\partial t} = \frac{i\hbar}{2m} \frac{\partial^2 \Psi}{\partial x^2} - \frac{i}{\hbar} V\Psi,$$

$$\frac{\partial \Psi^*}{\partial t} = -\frac{i\hbar}{2m} \frac{\partial^2 \Psi^*}{\partial x^2} + \frac{i}{\hbar} V\Psi^*.$$

So

$$\frac{\partial}{\partial t} |\Psi|^2 = \Psi^* \frac{\partial \Psi}{\partial t} + \frac{\partial \Psi^*}{\partial t} \Psi = \frac{i\hbar}{2m} \left(\Psi^* \frac{\partial^2 \Psi}{\partial x^2} - \frac{\partial^2 \Psi^*}{\partial x^2} \Psi \right)$$

$$= \frac{i\hbar}{2m} \frac{\partial}{\partial x} \left(\Psi^* \frac{\partial \Psi}{\partial x} - \frac{\partial \Psi^*}{\partial x} \Psi \right). \quad (10.25)$$

Hence,

$$\frac{d}{dt} \int_{-\infty}^{\infty} |\Psi(x,t)|^2 \, dx = \frac{i\hbar}{2m} \left(\Psi^* \frac{\partial \Psi}{\partial x} - \frac{\partial \Psi^*}{\partial x} \Psi \right) \Bigg|_{x \to -\infty}^{x \to +\infty} = 0,$$

since $\Psi, \Psi^* \to 0$ as $x \to \pm\infty$ in order that (10.23) can hold.   □

Relation (10.24) is a convenient property of the wave function. Once normalized, it is normalized for all t.

10.5.2   Momentum

As noted earlier, momentum can be viewed as an operator,

$$p\Psi = \left(\frac{\hbar}{i} \frac{\partial}{\partial x} \right) \Psi. \quad (10.26)$$

Another way to interpret this is as follows. If $\langle x \rangle$ is the expected value of the position x of the particle, we obtain

$$
\begin{aligned}
\frac{d\langle x \rangle}{dt} &= \int_{-\infty}^{\infty} x \frac{\partial}{\partial t}(\Psi^*\Psi)\, dx \\
&= \frac{i\hbar}{2m} \int_{-\infty}^{\infty} x \frac{\partial}{\partial x}\left(\Psi^*\frac{\partial\Psi}{\partial x} - \frac{\partial\Psi^*}{\partial x}\Psi\right) dx \qquad \text{(from (10.25))} \\
&= -\frac{i\hbar}{2m} \int_{-\infty}^{\infty} \left(\Psi^*\frac{\partial\Psi}{\partial x} - \frac{\partial\Psi^*}{\partial x}\Psi\right) dx \qquad \text{(integrating by parts)} \\
&= -\frac{i\hbar}{m} \int_{-\infty}^{\infty} \Psi^*\frac{\partial\Psi}{\partial x}\, dx \\
&= \frac{1}{m} \int_{-\infty}^{\infty} \Psi^*\left(\frac{\hbar}{i}\frac{\partial}{\partial x}\right)\Psi\, dx \\
&= \frac{1}{m} \int_{-\infty}^{\infty} \Psi^* p\Psi\, dx.
\end{aligned}
$$

In integrating by parts, we use the fact that $\Psi, \Psi^*, \partial\Psi/\partial x, \partial\Psi^*/\partial x \to 0$ as $x \to \pm\infty$, as in the proof of Proposition 10.1. We denote the final right-hand side by $\langle p \rangle/m$. Thus,

$$
\boxed{\langle p \rangle = m\frac{d\langle x \rangle}{dt} = \int_{-\infty}^{\infty} \Psi^* p\Psi\, dx.} \qquad (10.27)
$$

This is a noteworthy result. While we cannot prescribe momentum or position of a particle specifically in quantum mechanics due to Heisenberg's principle, the average (expected value) of p and x are related in precisely the way we would expect in the classical theory of particle dynamics; i.e., momentum is mass multiplied by velocity. It is important to realize that $\langle x \rangle$ is *not* the average of measurements of where the particle in question is (cf. Griffiths [21, p. 14]). Rather, $\langle x \rangle$ is the average of the positions of particles all in the same state Ψ. It is the ensemble average over systems in the identical state Ψ.

This relation of averages can be taken much further. If $V(x)$ is a potential, so that

$$
H(p, x) = \frac{p^2}{2m} + V(x),
$$

then

$$
\begin{aligned}
\frac{d \langle p \rangle}{dt} &= \frac{d}{dt} \int_{\mathbb{R}} \Psi^* p \Psi \; dx \\
&= \int_{\mathbb{R}} \frac{\partial \Psi^*}{\partial t} p \Psi \; dx + \int_{\mathbb{R}} \Psi^* p \frac{\partial \Psi}{\partial t} \; dx \\
&= \int_{\mathbb{R}} -\frac{1}{i\hbar} \Psi^* (Hp - pH) \Psi \; dx \qquad \text{(because } H = H^*) \\
&= \int_{\mathbb{R}} \left(\Psi^* V \frac{\partial \Psi}{\partial x} - \Psi^* \frac{\partial}{\partial x} (V\Psi) \right) dx \\
&= \int_{\mathbb{R}} \Psi^* \left(-\frac{\partial V}{\partial x} \right) \Psi \; dx \\
&= \left\langle -\frac{\partial V}{\partial x} \right\rangle .
\end{aligned}
$$

Thus,

$$
\frac{d \langle p \rangle}{dt} = \langle F \rangle, \tag{10.28}
$$

where $F = -\partial V/\partial x$ is the force acting on the particle. Clearly, these mean or expected values of p obey Newton's second law.

Bohr is credited with endowing such relationships between classical mechanics, in this case the dynamics of particles, with quantum theory, with the status of a basic principle called the *correspondence principle*. Messiah [37, p. 29] describes the principle as one asserting that "*Classical Theory* is macroscopically correct; that is to say, it accounts for phenomena in the limit where quantum discontinuities may be considered infinitely small; in all these limiting cases, the predictions of the exact theory must coincide with those of the Classical Theory." By "exact theory", he refers to quantum mechanics. He goes on to say, equivalently, that "Quantum Theory must approach Classical Theory asymptotically in the limit of large quantum numbers." One might add that were this not the case, the rich and very successful macroscopic theories of continuum mechanics and electromagnetic field theory developed earlier would be much less important.

10.5.3   Wave Packets and Fourier Transforms

Consider again the case of rectilinear motion of a free particle x in a vacuum. The wave function is

$$\Psi(x,t) = \psi(x)e^{-i\omega t}$$
$$= \psi(x)e^{-iEt/\hbar}, \tag{10.29}$$

and $\psi(x)$ satisfies

$$-\frac{\hbar^2}{2m}\frac{d^2\psi}{dx^2} = E\psi. \tag{10.30}$$

The solution is

$$\Psi(x,t) = Ae^{ik(x-\hbar kt/2m)}, \tag{10.31}$$

where

$$k = \pm\sqrt{\frac{2mE}{\hbar^2}}. \tag{10.32}$$

Clearly, $E = \hbar^2 k^2/2m = 2\pi^2\hbar^2/m\lambda^2$, so that the energy increases as the wavelength λ decreases.

This particular solution is not normalizable ($\int_{\mathbb{R}} \Psi^*\Psi dx \to \infty$), so that a free particle cannot exist in a stationary state. The spectrum in this case is continuous, and the solution is of the form

$$\Psi(x,t) = \frac{1}{\sqrt{2\pi}}\int_{-\infty}^{\infty} \varphi(k)e^{i(kx-\hbar k^2 t/2m)}dk, \tag{10.33}$$

$$\Psi(x,0) = \frac{1}{\sqrt{2\pi}}\int_{-\infty}^{\infty} \varphi(k)e^{ikx}dx, \tag{10.34}$$

which implies that

$$\varphi(k) = \frac{1}{\sqrt{2\pi}}\int_{-\infty}^{\infty} \Psi(x,0)e^{-ikx}dx. \tag{10.35}$$

Equation (10.33) characterizes the wave as a sum of *wave packets*. It is a sinusoidal function modulated by the function φ. Here φ is the Fourier transform of the wave function Ψ, and the initial value $\Psi(x,0)$ of Ψ is the inverse Fourier transform of φ. Instead of a particle velocity as in classical mechanics, we have a *group velocity* of the envelope of wavelets.

10.6   The Wave–Momentum Duality

In classical mechanics, the dynamical state of a particle is defined at every instant of time by specifying its position $\mathbf{x}$ and its momentum $\mathbf{p}$. In quantum mechanics, one can only define the probability of finding the particle in a given region when one carries out a measurement of the position. Similarly, one cannot define the momentum of a particle; one can only hope to define the probability of determining the momentum in a given region of *momentum space* when carrying out a measure of momentum.

To define the probability of finding momentum $\mathbf{p}$ in a volume $(\mathbf{p}, \mathbf{p} + d\mathbf{p})$, we consider, for fixed time t, the Fourier transform $\Phi(\mathbf{p})$ of the wave function Ψ:

$$
\Phi(\mathbf{p}) = \frac{1}{(2\pi\hbar)^{n/2}} \int_{\mathbb{R}^n} \Psi(\mathbf{x}) e^{-i\mathbf{p}\cdot\mathbf{x}/\hbar} dx,
$$

$$
\Psi(\mathbf{x}) = \frac{1}{(2\pi\hbar)^{n/2}} \int_{\mathbb{R}^n} \Phi(\mathbf{p}) e^{i\mathbf{p}\cdot\mathbf{x}/\hbar} dp,
$$

(10.36)

with $dx = dx_1 \cdots dx_N$ and $dp = dp_1 \cdots dp_N$. Thus, the wave function can be viewed as a linear combination of waves $\exp(i\mathbf{p} \cdot \mathbf{x}/\hbar)$ of momentum $\mathbf{p}$ with coefficients $(2\pi\hbar)^{-n/2}\Phi(\mathbf{p})$. The probability of finding a momentum $\mathbf{p}$ in the volume $(\mathbf{p}, \mathbf{p} + d\mathbf{p})$ is

$$
\pi(\mathbf{p}) = \Phi^*(\mathbf{p})\, \Phi(\mathbf{p}),
$$

and we must have

$$
\int_{\mathbb{R}^n} \Phi^*(\mathbf{p})\, \Phi(\mathbf{p})\, dp = 1. \tag{10.37}
$$

Thus, the Fourier transform $\mathcal{F} : L^2(\mathbb{R}^n) \to L^2(\mathbb{R}^n)$ establishes a one-to-one correspondence between the wave function and the momentum wave function. Equation (10.37) follows from Plancherel's identity,

$$
\langle \Psi | \Psi \rangle = \|\Psi\|^2 = \langle \mathcal{F}(\Psi) | \mathcal{F}(\Psi) \rangle = \langle \Phi | \Phi \rangle = \|\Phi\|^2. \tag{10.38}
$$

Here $\langle \,|\, \rangle = \langle \,,\, \rangle$ denotes an inner product on $L^2(\mathbb{R}^n)$, discussed in more detail in the next chapter.

The interpretation of the probability densities associated with Φ and Ψ is important. When carrying out a measurement on either position or momentum, neither can be determined with precision. The predictions of the probabilities of position and momentum are understood to mean that a very large number M of equivalent systems with the same wave function Ψ are considered. A position measurement on each of them gives the probability density $\Psi^*(\mathbf{x})\Psi(\mathbf{x})$ of results in the limit as M approaches infinity. Similarly, $\Phi^*(\mathbf{x})\Phi(\mathbf{x})$ gives the probability density of results of measuring the momentum.

10.7 Appendix: A Brief Review of Probability Densities

We have seen that the solution Ψ of Schrödinger's equation is such that $\Psi^*\Psi = |\Psi|^2$ is a probability density function, ρ, defining the probability that the particle in question is in the interval $(x, x + dx]$ is $\rho\,dx$. It is appropriate to review briefly the essential ideas of probability theory.

We begin with the idea of a set Ω of samples of objects called the sample set (or space) and the class $U = U(\Omega)$ of subsets of Ω which has the property that unions, intersections and complements of subsets in U also belong to U, including Ω and the null set $\emptyset$ (properties needed to make sense of the notion of probability). The class U is called a σ-algebra on Ω, and the sets in U are called *events*. Next, we introduce a map $\mathbb{P}$ from U onto the unit interval $[0, 1]$ which has the properties

$$\mathbb{P}(A \cup B) = \mathbb{P}(A) + \mathbb{P}(B), \qquad A \cap B = \emptyset, \text{ or}$$
$$\leq \mathbb{P}(A) + \mathbb{P}(B), \qquad \forall A, B \in U,$$
$$\mathbb{P}(\Omega) = 1,$$
$$\mathbb{P}(\emptyset) = 0.$$

These properties qualify $\mathbb{P}$ to be a *measure* on U. The number $\mathbb{P}(A)$, $A \in U$, is the *probability* of the event A. The triple $(\Omega, U, \mathbb{P})$ is called a *probability space*.

Example Suppose we have a single die (half a pair of dice), which is a cube with six numbers, one on each side, represented by the usual dots:

⊡, ⊡, ⊡, ⊞, ⊠, ⊟. These six possibilities form the set Ω. We are going to roll the die on a table top and see what number ends facing up when the die comes to rest. The space U of possible events is the power set 2^{Ω}: the set of all subsets of Ω, including Ω itself and $\emptyset$. (Since Ω contains six members, U contains $2^6 = 64$ possible events.)

Now a probability $\mathbb{P}$ on U is a *function* from U into $[0, 1]$. What is a function? Basically, it is a rule that establishes a relation between elements of U, the events, and numbers in $[0, 1]$ such that

1. every event A has an image $\mathbb{P}(A)$ in $[0, 1]$, and

2. for each such A, one and only one number $a \in [0, 1]$ is an image $\mathbb{P}(A)$.

So, to define a probability on U we need a rule. Here are some examples:

- What is the probability that the upward face is 4 (⊞)? The standard answer is determined by the *frequency* of events, and the "odds" of 4 is one of six, so $\mathbb{P}(\{4\}) = 1/6$.

- What is the probability of the upward face to be an even number? Then, since three out of the six are even, $\mathbb{P}(A) = 3/6 = 1/2$.

- What is the probability the number is 7? Since $\{7\} \notin U$, we have $\mathbb{P}(\{7\}) = 0$.

- What is the probability the upward face is between 1 and 6? $\mathbb{P}(\Omega) = 1$.

$\square$

Since probability spaces are not readily observable, we work with them using the idea of a real variable. A real random variable X is a map from Ω into $\mathbb{R}$ such that, for any open set $B \subset \mathbb{R}$, we have $X^{-1}(B) \in U$. Basically, the inverse image of a random variable maps an open set $B \in \mathbb{R}$, which itself belongs to a σ-algebra $\mathcal{B}$ on $\mathbb{R}$ called the Borel sigma algebra on $\mathbb{R}$, into events $A \in U$. So it makes sense to speak of the probability $\mathbb{P}(X^{-1}(B))$, usually written

$$\mathbb{P}(X \in B) \equiv \mathbb{P}(\{\omega : X(\omega) \in B\}).$$

This is an awkward notation, so we seek to replace it by some equivalent idea involving real-valued functions. This is accomplished by introducing the *distribution function* $F_X : \mathbb{R} \to [0, 1]$ for the random variable X defined by

$$F_X(x) = \mathbb{P}(X \leq x)$$

$\forall A \in U, \; x \in \mathbb{R}$. This is read: Given a random variable X, the probability distribution function for X takes on values $F_X(x)$ at each $x \in \mathbb{R}$ equal to the probability that $X(A) \leq x$, for any event $A \in U$.

It is generally possible to make a further characterization of X by introducing a function $\rho = \rho(x)$ such that

$$F_X(x) = \int_{-\infty}^{x} \rho(y) \, dy.$$

The function ρ is called the *probability density* for X. If such a function exists, it has the properties

$$\mathbb{P}(X \in B) = \int_{B} \rho(x) \, dx, \qquad B \in \mathcal{B},$$

$$\int_{-\infty}^{\infty} \rho(x) \, dx = 1.$$

Given a random variable X with probability density ρ, the *mean* or *expected value* of X is denoted $\langle x \rangle$ and is defined by

$$\langle x \rangle = \int_{-\infty}^{\infty} x \, \rho(x) \, dx.$$

The central moments are

$$\mu_k = \int_{-\infty}^{\infty} \left(x - \langle x \rangle \right)^{k} \rho(x) \, dx, \qquad k = 1, 2, \ldots.$$

In particular, the *variance* is defined by

$$\sigma_X^2 = \mu_2 = \int_{-\infty}^{\infty} \left(x - \langle x \rangle \right)^{2} \rho(x) \, dx,$$

and the *standard deviation* is $\sigma_X = \sqrt{\sigma_X^2} = \sqrt{\mu_2}$. We easily confirm that

$$\sigma_X^2 = \langle x^2 \rangle - \langle x \rangle^2.$$

DYNAMICAL VARIABLES AND OBSERVABLES IN QUANTUM MECHANICS: THE MATHEMATICAL FORMALISM

11.1 Introductory Remarks

We have seen that the solutions of Schrödinger's equations are wave functions $\Psi = \Psi(x, t)$ that have the property that $|\Psi|^2 = \Psi^*\Psi$ is the probability density function giving the probability that the elementary particle under study is at position x at time t (actually that the particle is in the volume between x and $x + dx$). Moreover, the knowledge of the wave function Ψ (or equivalently, the momentum wave function $\Phi = \Phi(p, t)$) determines completely the dynamical state of the quantum system. We shall now build on these ideas and develop the appropriate mathematical setting for an operator theoretic framework for quantum mechanics that brings classical tools and concepts to the theory.

Everything we derive is applicable to functions defined on $\mathbb{R}^d$ ($d = 1, 2, 3$), but virtually all of the results can be demonstrated without loss of generality on $\mathbb{R}$. The notation for the spatial coordinate

An Introduction to Mathematical Modeling: A Course in Mechanics, First Edition. By J. Tinsley Oden
© 2011 John Wiley & Sons, Inc. Published 2011 by John Wiley & Sons, Inc.

x will be used interchangeably with **q** (or x with q in $\mathbb{R}$), as **q** is the classical notation for a generalized coordinate in "phase space" where coordinate–momentum pairs $(\mathbf{q}, \mathbf{p}) = (\mathbf{q}_1, \mathbf{q}_2, \ldots, \mathbf{q}_N, \mathbf{p}_1, \mathbf{p}_2, \ldots, \mathbf{p}_N)$ define dynamical states. We will frequently treat dynamical variables, such as momentum p, as operators, and when it is important to emphasize the operator character of the result, we will affix a tilde to the symbol (i.e. for momentum p, the associated operator is $\tilde{p} = -i\hbar \frac{d}{dq}$). Thus, a function $F = F(\mathbf{q}, \mathbf{p})$ is associated with an operator $\tilde{F}(\mathbf{q}_1, \mathbf{q}_2, \ldots, \mathbf{q}_N, -i\hbar \frac{\partial}{\partial \mathbf{q}_1}, -i\hbar \frac{\partial}{\partial \mathbf{q}_2}, \ldots, -i\hbar \frac{\partial}{\partial \mathbf{q}_N})$ etc. The coordinates $(\mathbf{q}_1, \mathbf{q}_2, \ldots, \mathbf{q}_N)$ will hereafter be understood to be Cartesian coordinates because the operator notation must represent dynamical quantities in a way that is invariant under a change (e.g. a rotation) of the coordinate axes. Indeed, while ordinary multiplication of functions is commutative, the corresponding operators may not commute, so the q_i are interpreted as Cartesian coordinates to avoid ambiguity.

11.2 The Hilbert Spaces $L^2(\mathbb{R})$ (or $L^2(\mathbb{R}^d)$) and $H^1(\mathbb{R})$ (or $H^1(\mathbb{R}^d)$)

Since the wave function must have the property that $|\Psi|^2$ is integrable over $\mathbb{R}$ (Ψ is "square-integrable"), it must belong to the following space:

$L^2(\mathbb{R})$ *(or $L^2(\mathbb{R}^d)$) is the space of equivalence classes $[u]$ of measurable complex-valued functions $u : \mathbb{R} \to \mathbb{C}$ equal almost everywhere on $\mathbb{R}$ (or $\mathbb{R}^d$) ($v \in [u] \Rightarrow v = u$ everywhere except on sets of measure zero) such that $|u|^2 = u^*u$ is Lebesgue integrable on $\mathbb{R}$ (or $\mathbb{R}^d$).*

Thus, $u \in L^2(\mathbb{R})$ implies that u represents an equivalence class $[u]$ of functions equal almost everywhere on $\mathbb{R}$ such that $\int_{\mathbb{R}} u^2 dx < \infty$.

The space $L^2(\mathbb{R})$ is a complete inner product space with the inner product of two functions ψ and φ in $L^2(\mathbb{R})$ defined by

$$\langle \psi, \varphi \rangle = \int_{\mathbb{R}} \psi^* \varphi \, dx, \tag{11.1}$$

where ψ^* is the complex conjugate of ψ. By a "complete" space, we mean one in which every infinite sequence of elements $\{u_j\}_{j=1}^{\infty}$, the entries of which get closer together in the metric distance of the space as $j \to \infty$ (i.e., $\|u_j - u_k\| \to 0$ as $j, k \to \infty$), always converges to an element u of the space. Such sequences are called *Cauchy sequences*. Complete inner product spaces are called *Hilbert spaces*. Thus $L^2(\mathbb{R})$ is a Hilbert space. The associated norm on $L^2(\mathbb{R})$ is then

$$\|\psi\| = \sqrt{\langle \psi, \psi \rangle}. \tag{11.2}$$

It can be shown that $L^2(\mathbb{R})$ is *reflexive* (in particular, by the Riesz theorem, for every continuous linear functional f in the dual $(L^2(\mathbb{R}))'$, there is a unique $u_f \in L^2(\mathbb{R})$ such that $f(v) = \langle v, u_f \rangle$ and $\|f\|_{(L^2(\mathbb{R}))'} = \|u_f\|$. Also, $L^2(\mathbb{R})$ is *separable*, meaning that it contains countable everywhere dense sets. In other words, for any $u \in L^2(\mathbb{R})$ and any $\varepsilon > 0$, there exists an infinite sequence of functions $\{u_n\}_{n=1}^{\infty}$ in $L^2(\mathbb{R})$ and an integer $M > 0$ such that $\|u_n - u\| < \varepsilon$ for all $n > M$. We shall demonstrate how such countable sets can be computed given any u in the next section.

We are also interested in spaces of functions that have partial derivatives in $L^2(\mathbb{R}^d)$. This is the Sobolev space,

$$\boxed{H^1(\mathbb{R}^d) = \left\{ v \in L^2(\mathbb{R}^d) \text{ and } \frac{\partial v}{\partial x_i} \in L^2(\mathbb{R}^d);\ i = 1, 2, \dots, d \right\}.}$$

$$\tag{11.3}$$

This is also a Hilbert space with inner product,

$$\langle \psi, \varphi \rangle_1 = \int_{\mathbb{R}^d} \left(\nabla \psi^* \cdot \nabla \varphi + \psi^* \varphi \right) d^d x, \tag{11.4}$$

where

$$\nabla \psi^* \cdot \nabla \varphi = \sum_{i=1}^{d} \frac{\partial \psi^*}{\partial x_i} \frac{\partial \varphi}{\partial x_i},$$

and $d^d x = dx_1 \, dx_2 \cdots dx_d$. When we write the $L^2(\mathbb{R}^d)$-inner product of a function with $H\psi$, we regard the derivatives in H in a generalized sense. So,

$$\langle \varphi, H\psi \rangle = \int_{\mathbb{R}^d} \left(\frac{\hbar^2}{2m} \nabla \varphi^* \cdot \nabla \psi + V \, \varphi^* \, \psi \right) d^d x, \tag{11.5}$$

which is well-defined for $H^1(\mathbb{R}^d)$ functions (for smooth V). If φ and ψ are smooth enough, an integration by parts of (11.5) gives

$$\begin{aligned} \langle \varphi, H\psi \rangle &= \int_{\mathbb{R}^d} \varphi^* \left(-\frac{\hbar^2}{2m} \Delta \psi + V \, \psi \right) d^d x \\ &= \int_{\mathbb{R}^d} \varphi^* \, H\psi \, d^d x. \end{aligned}$$

For functions with partial derivatives of order $m \geq 0$ in $L^2(\mathbb{R}^d)$, we analogously define spaces $H^m(\mathbb{R}^d)$ with inner products defined in an analogous way.

11.3 Dynamical Variables and Hermitian Operators

A *dynamical variable* is some physical feature of the quantum system that depends upon the physical state of the system. In general, we adopt the convention that the state is described by the position coordinates $\mathbf{q}$ and the momentum $\mathbf{p}$ (or for a system with N particles, $\mathbf{q}_1, \mathbf{q}_2, \ldots, \mathbf{q}_N; \mathbf{p}_1, \mathbf{p}_2, \ldots, \mathbf{p}_N$). Thus, a dynamical variable is a function

$$Q = Q(\mathbf{q}_1, \mathbf{q}_2, \ldots, \mathbf{q}_N; \mathbf{p}_1, \mathbf{p}_2, \ldots, \mathbf{p}_N).$$

Since the momentum components $\mathbf{p}_j$ can be associated with the operators $\tilde{\mathbf{p}}_j = -i\hbar \, \partial/\partial \mathbf{q}_j$, dynamical variables likewise characterize operators

$$\tilde{Q}\left(\mathbf{q}_1, \mathbf{q}_2, \ldots, \mathbf{q}_N; -i\hbar \frac{\partial}{\partial \mathbf{q}_1}, -i\hbar \frac{\partial}{\partial \mathbf{q}_2}, \ldots, -i\hbar \frac{\partial}{\partial \mathbf{q}_N} \right).$$

Analogously, one could define an operator

$$\hat{Q}\left(-i\hbar\frac{\partial}{\partial \mathbf{p}_1}, -i\hbar\frac{\partial}{\partial \mathbf{p}_2}, \ldots, -i\hbar\frac{\partial}{\partial \mathbf{p}_N}; \mathbf{p}_1, \mathbf{p}_2, \ldots, \mathbf{p}_N\right).$$

The *expected value* or *mean* of a dynamical variable Q for quantum state Ψ is denoted $\langle Q \rangle$ and is defined as

$$\langle Q \rangle = \int_{\mathbb{R}} \Psi^* \tilde{Q} \Psi \, dq \Big/ \int_{\mathbb{R}} \Psi^* \Psi \, dq$$
$$= \frac{\langle \Psi, \tilde{Q}\Psi \rangle}{\langle \Psi, \Psi \rangle}$$
$$= \langle \Psi, \tilde{Q}\Psi \rangle, \qquad (11.6)$$

if $\langle \Psi, \Psi \rangle = 1$.

Any operator $A : L^2(\mathbb{R}) \to L^2(\mathbb{R})$ is said to be *Hermitian* if

$$\langle \psi, A\varphi \rangle = \langle A\psi, \varphi \rangle \quad \forall \psi, \varphi \in L^2(\mathbb{R}). \qquad (11.7)$$

In other words, a Hermitian operator has the property

$$\int_{\mathbb{R}} \psi^* A\varphi \, dq = \int_{\mathbb{R}} (A\psi)^* \varphi \, dq,$$

for arbitrary ψ and φ in $L^2(\mathbb{R})$.

Any operator Q that corresponds to a genuine physically *observable* feature of a quantum system must be such that its expected value $\langle Q \rangle$ is real; i.e.,

$$\langle Q \rangle = \langle \Psi, \tilde{Q}\Psi \rangle$$

is real. Therefore, for such Q we must have

$$\langle \psi, \tilde{Q}\varphi \rangle = \langle \tilde{Q}\psi, \varphi \rangle \quad \forall \psi, \varphi \in L^2(\mathbb{R}).$$

In other words:

> *Every observable must be characterized by a Hermitian operator on $L^2(\mathbb{R})$ (or $L^2(\mathbb{R}^d)$).*

(Technically, we reserve the word "observable" to describe dynamical variables defined by Hermitian operators which have a complete orthonormal set of eigenfunctions, a property taken up in the next section; see Theorem 11.B.)

We note that the *variance* of a dynamical variable is defined as

$$\sigma_Q^2 = \langle Q^2 \rangle - \langle Q \rangle^2. \tag{11.8}$$

The dynamical variable Q takes on the numerical value $\langle Q \rangle$ with certainty if and only if $\sigma_Q^2 = 0$. This observation leads us to a remarkable property of those dynamical variables that can actually be measured in quantum systems with absolute certainty, the so-called *observables*. For a Hermitian operator $\tilde{Q}$, we have

$$\begin{aligned}
\sigma_Q^2 &= \langle Q^2 \rangle - \langle Q \rangle^2 \\
&= \langle Q \rangle^2 - 2\langle Q \rangle \langle Q \rangle + \langle Q \rangle \\
&= \langle (Q - \langle Q \rangle)^2 \rangle \\
&= \langle (\tilde{Q} - \langle Q \rangle)\Psi, (\tilde{Q} - \langle Q \rangle)\Psi \rangle \\
&= \left\| (\tilde{Q} - \langle Q \rangle)\Psi \right\|^2.
\end{aligned} \tag{11.9}$$

Thus, if $\sigma_Q^2 = 0$, we have

$$\boxed{\tilde{Q}\Psi = \langle Q \rangle \Psi.} \tag{11.10}$$

This is recognized as the *eigenvalue problem for the Hermitian operator* $\tilde{Q}$, with eigenvalue–eigenfunction pair $(\langle Q \rangle, \Psi)$. Therefore, the "fluctuations" (variances) of the dynamical variable Q from its mean $\langle Q \rangle$ vanish for states Ψ which are eigenfunctions of the operator $\tilde{Q}$, and the expected value $\langle Q \rangle$ is an eigenvalue of $\tilde{Q}$. We summarize this finding in the following theorem:

Theorem 11.A *The dynamical variable Q of a quantum system possesses with certainty (with probability 1) the well-defined value $\langle Q \rangle$ if and only if the dynamical state of the system is represented by an eigenfunction Ψ of the Hermitian operator $\tilde{Q} : L^2(\mathbb{R}) \to L^2(\mathbb{R})$ associated with Q; moreover, the expected value $\langle Q \rangle$ is an eigenvalue of $\tilde{Q}$, and Q is then an observable feature of the system.*

A canonical example is the time-independent Schrödinger equation,

$$\tilde{H}\psi = E\psi.$$

Thus, as we have seen earlier, the determinate energy states are eigenvalues of the Hamiltonian.

Theorem 11.A establishes the connection of quantum mechanics with the spectral theory of Hermitian operators. We explore this theory in more detail for the case of a discrete spectrum.

11.4 Spectral Theory of Hermitian Operators: The Discrete Spectrum

Returning to the eigenvalue problem (11.10), let us consider the case in which there exists a countable but infinite set of eigenvalues λ_k and eigenfunctions $\varphi_k \in L^2(\mathbb{R})$, $k = 1, 2, \ldots$ for the operator $\tilde{Q}$. In this case, $\tilde{Q}$ is said to have a *discrete spectrum*. The basic properties of the system in this case are covered or derived from the following theorem:

Theorem 11.B *Let (λ_k, φ_k), $k = 1, 2, \ldots$, denote a countable sequence of eigenvalue–eigenfunction pairs for the compact Hermitian operator $\tilde{Q} : L^2(\mathbb{R}) \to L^2(\mathbb{R})$; i.e.,*

$$\tilde{Q}\varphi_k = \lambda_k \varphi_k, \quad k = 1, 2, \ldots. \tag{11.11}$$

Then

1. *If φ_k is an eigenfunction, so also is $c\,\varphi_k$, c being any constant.*

2. *If $\varphi_k^{(1)}, \varphi_k^{(2)}, \ldots, \varphi_k^{(M)}$ are M eigenfunctions corresponding to the same eigenvalue λ_k, then any linear combination of these eigenfunctions is an eigenfunction corresponding to λ_k.*

3. *The eigenvalues are real.*

4. *The eigenfunctions φ_k can be used to construct an orthonormal set—i.e., a set of eigenfunctions of $\tilde{Q}$ such that*

$$\langle \varphi_k, \varphi_m \rangle = \delta_{km} = \begin{cases} 1 & \text{if } k = m, \\ 0 & \text{if } k \neq m. \end{cases} \tag{11.12}$$

5. *Any state $\Psi \in L^2(\mathbb{R})$ can be represented as a series,*

$$\Psi(q) = \sum_{k=1}^{\infty} c_k \varphi_k(q), \tag{11.13}$$

where

$$c_k = \langle \varphi_k, \Psi \rangle, \tag{11.14}$$

and by (11.13) we mean

$$\lim_{m \to \infty} \left\| \Psi - \sum_{k=1}^{m} c_k \varphi_k \right\| = 0. \tag{11.15}$$

Proof Parts 1 and 2 are trivial. In property 2, the eigenfunctions are said to have a *degeneracy* of order M. To show 3, we take the inner product of both sides of (11.11) by φ_k and obtain the number

$$\lambda_k = \langle \varphi_k, \tilde{Q}\varphi_k \rangle / \langle \varphi_k, \varphi_k \rangle,$$

which is real, because $\tilde{Q}$ is Hermitian.

To show 4, consider

$$\tilde{Q}\varphi_k = \lambda_k \varphi_k \quad \text{and} \quad \tilde{Q}\varphi_m = \lambda_m \varphi_m, \quad m \neq k.$$

Then,

$$\langle \varphi_m, \tilde{Q}\varphi_k \rangle = \lambda_k \langle \varphi_m, \varphi_k \rangle = \langle \tilde{Q}\varphi_m, \varphi_k \rangle = \lambda_m \langle \varphi_m, \varphi_k \rangle.$$

Thus, $(\lambda_k - \lambda_m)\langle \varphi_m, \varphi_k \rangle = 0$ for $m \neq k$, $\lambda_k \neq \lambda_m$.

Finally, to show 5, the eigenfunctions φ_k are assumed to be normalized:

$$\langle \varphi_k, \varphi_k \rangle = 1, \quad k = 1, 2, \ldots.$$

Thus they form an orthonormal basis for $L^2(\mathbb{R})$. Equation (11.13) is then just the Fourier representation of Ψ with respect to this basis. Equation (11.14) follows by simply computing $\langle \sum_m c_m \varphi_m, \varphi_k \rangle$ and using Eq. (11.12). $\qquad\square$

Various functions of the operator $\tilde{Q}$ can likewise be given a spectral representation. For instance,

$$\text{if} \quad \tilde{Q}\varphi_k = \lambda_k \varphi_k, \quad \text{then} \quad \tilde{Q}^2 \varphi_k = \tilde{Q}(\tilde{Q}\varphi_k)$$
$$= \tilde{Q}\lambda_k \varphi_k$$
$$= \lambda_k \tilde{Q}\varphi_k$$
$$= \lambda_k^2 \varphi_k,$$

and, in general,

$$(\tilde{Q})^r \varphi_k = \lambda_k^r \varphi_k.$$

Symbolically,

$$(\sin \tilde{Q})\varphi_k = (\tilde{Q} - \frac{1}{3!}\tilde{Q}^3 + \cdots)\varphi_k$$
$$= (\lambda_k - \frac{1}{3!}\lambda_k^3 + \cdots)\varphi_k$$
$$= \sin(\lambda_k)\varphi_k,$$

etc. Suppose

$$u = \sum_{k=1}^{\infty} b_k \varphi_k,$$

then

$$(\sin \tilde{Q})u = \sum_{k=1}^{\infty} b_k (\sin \lambda_k)\,\varphi_k.$$

In general, if $\tilde{Q}$ is a Hermitian operator with discrete (nondegenerate, for simplicity in notation) eigenvalue–eigenfunction pairs (λ_k, φ_k) and F is any smooth function on $\mathbb{R}$, we may write

$$F(\tilde{Q})u = \sum_{k=1}^{\infty} b_k F(\lambda_k)\,\varphi_k. \tag{11.16}$$

For $\tilde{Q} : L^2(\mathbb{R}) \to L^2(\mathbb{R})$, (11.16) is meaningful if the series converges; i.e. if the sequence of real numbers,

$$\mu_n = \sum_{k=1}^{n} |b_k F(\lambda_k)|^2,$$

converges in $\mathbb{R}$. For

$$F(\tilde{Q}) = e^{i\xi\tilde{Q}}, \quad \xi \in \mathbb{R},$$

this series always converges. In particular,

$$e^{i\xi\tilde{Q}}\Psi = \sum_{k=1}^{\infty} c_k e^{i\xi\tilde{Q}} \varphi_k$$

$$= \sum_{k=1}^{\infty} c_k e^{i\xi\lambda_k} \varphi_k,$$

and $\left|c_k e^{i\xi\lambda_k}\right|^2 = |c_k|^2 \to 0$ as $k \to \infty$.

11.5 Observables and Statistical Distributions

The statistical distribution of a quantity Q associated with a quantum dynamical system is established by its *characteristic function* $a : \mathbb{R} \to \mathbb{R}$. In general, the characteristic function is, to within a constant, the Fourier transform of the probability density function ϱ associated with a random variable X:

$$a(\xi) = \int_{-\infty}^{\infty} e^{i\xi x} \varrho(x) \, dx, \qquad (11.17)$$

$\varrho(x)$ being the probability of finding X in $(x, x + dx)$. Thus, $a(\xi)$ is the expected value of $e^{i\xi x}$:

$$a(\xi) = \langle e^{i\xi x} \rangle. \qquad (11.18)$$

The generalization of this case to a Hermitian operator $\tilde{Q}$ on $L^2(\mathbb{R})$ reveals, as we shall see, an important property of the coefficients c_k of (11.14). First, we note that if the random variable X can only assume discrete values x_k, $k = 1, 2, \ldots$, and if $\varrho_1, \varrho_2, \ldots$ are the probabilities of these values, then

$$a(\xi) = \sum_{k=\infty}^{\infty} \varrho_k e^{i\xi x_k}, \qquad (11.19)$$

with $\sum_k \varrho_k = 1$.

Returning now to quantum dynamics, as noted earlier, any dynamical variable Q of a quantum system characterized by a Hermitian operator $\tilde{Q}$ on $L^2(\mathbb{R})$ with a discrete spectrum of eigenvalues λ_k is an observable, although observables may be described by operators with a continuous spectrum, noted in the next section. The reason for that term is clear from Theorem 11.A, but the role of the characteristic function $a(\xi)$ gives further information. For the wave function Ψ representing the dynamical state of a quantum system, we have

$$a(\xi) = \frac{\langle \Psi, e^{i\xi \tilde{Q}} \Psi \rangle}{\langle \Psi, \Psi \rangle}$$

$$= \sum_{k=\infty}^{\infty} \varrho_k e^{i\xi \lambda_k}, \qquad (11.20)$$

where now

$$\varrho_k = |c_k|^2 / \langle \Psi, \Psi \rangle$$
$$= |\langle \varphi_k, \Psi \rangle|^2 / \langle \Psi, \Psi \rangle. \qquad (11.21)$$

Comparing (11.19) and (11.20), we arrive at the following theorem:

Theorem 11.C *Let Q be an observable of a quantum dynamical system with the discrete spectrum of eigenvalues $\{\lambda_k\}_{k=1}^{\infty}$. Then the probability that Q takes on the value λ_k is*

$$\varrho_k = |c_k|^2 = |\langle \varphi_k, \Psi \rangle|^2,$$

where φ_k is the eigenfunction corresponding to λ_k and Ψ is the normalized wave function, $\langle \Psi, \Psi \rangle = 1$.

Thus, the only values Q can assume are its eigenvalues $\lambda_1, \lambda_2, \ldots$, and the probability that Q takes on the value λ_k is $|c_k|^2 = |\langle \varphi_k, \Psi \rangle|^2$. We observe that the discrete probabilities ϱ_k satisfy

$$1 = \langle \Psi, \Psi \rangle = \left\langle \sum_{k=1}^{\infty} c_k \varphi_k, \sum_{m=1}^{\infty} c_m \varphi_m \right\rangle$$

$$= \sum_k \sum_m c_k^* c_m \langle \varphi_k, \varphi_m \rangle$$

$$= \sum_{k=1}^{\infty} |c_k|^2 = \sum_{k=1}^{\infty} \varrho_k,$$

as required.

11.6 The Continuous Spectrum

Not all dynamical variables (Hermitian operators) have a discrete spectrum; i.e., not all are observables in quantum dynamical systems. For example, the momentum operator $p = -i\hbar \, d/dq$ is Hermitian but does not have a discrete spectrum and, therefore, is not an observable (in keeping with our earlier view of the uncertainty principle for position and momentum). In particular, the eigenvalue problem for p is (cf. Messiah [37, pp. 179–190])

$$p\,U(p',q) = p'\,U(p',q), \tag{11.22}$$

where $U(p',q)$ is the eigenfunction of p with an eigenvalue $\lambda = p'$, a continuous function of the variable p'; i.e., the spectrum of $p = -i\hbar\frac{d}{dq}$ is continuous. We easily verify that

$$U(p',q) = \frac{1}{\sqrt{2\pi\hbar}} e^{ip'q/\hbar}. \tag{11.23}$$

By analogy with the Fourier series representation (11.13) of the dynamical state ψ for discrete spectra, we now have the Fourier transform representation of the state for the case of a continuous spectrum:

$$\psi(q) = \frac{1}{\sqrt{2\pi\hbar}} \int_{\mathbb{R}} \varphi(p')e^{ip'q/\hbar}dp'. \tag{11.24}$$

In analogy with (11.14), we have

$$\varphi(p') = \langle U(p',q), \psi(q) \rangle. \tag{11.25}$$

The analogue arguments cannot be carried further. Indeed, the function (11.23) is not in $L^2(\mathbb{R})$, and a more general setting must be constructed to put the continuous spectrum case in the proper mathematical framework.

11.7 The Generalized Uncertainty Principle for Dynamical Variables

Recall that for any dynamical variable characterized by a Hermitian operator Q, its expected value in the state Ψ is

$$\langle Q \rangle = \langle \Psi, \tilde{Q}\Psi \rangle,$$

where $\tilde{Q}$ is the operator associated with Q ($Q(x, p, t) = \tilde{Q}(x, \frac{\hbar}{i}\frac{\partial}{\partial x}, t)$), and $\langle \Psi, \Psi \rangle = 1$. Recall that its variance is

$$\sigma_Q^2 = \langle (Q - \langle Q \rangle)^2 \rangle = \langle \Psi, (\tilde{Q} - \langle Q \rangle)^2 \Psi \rangle. \tag{11.26}$$

For any other such operator M, we have

$$\sigma_M^2 = \langle \Psi, (\tilde{M} - \langle M \rangle)^2 \Psi \rangle$$
$$= \langle (\tilde{M} - \langle M \rangle)\Psi, (\tilde{M} - \langle M \rangle)\Psi \rangle.$$

Noting that for any complex number z, $|z|^2 \geq (\mathrm{Im}(z))^2 = \left(\frac{1}{2i}(z - z^*)\right)^2$, it is an algebraic exercise (see exercise 1 in Set II.2) to show that

$$\boxed{\sigma_Q^2 \sigma_M^2 \geq \left(\frac{1}{2i}\left\langle [\tilde{Q}, \tilde{M}]\right\rangle\right)^2,} \tag{11.27}$$

where $[\tilde{Q}, \tilde{M}]$ is the *commutator* of the operators Q and M:

$$[\tilde{Q}, \tilde{M}] = \tilde{Q}\tilde{M} - \tilde{M}\tilde{Q}. \tag{11.28}$$

The generalized Heisenberg uncertainty principle is this: For any pair of incompatible observables (those for which the operators do not commute, or $[\tilde{Q}, \tilde{M}] \neq 0$), the condition (11.27) holds. We will show that such incompatible operators cannot share common eigenfunctions. The presence of the i in (11.27) does not render the right-hand side negative, as $[\tilde{Q}, \tilde{M}]$ may also involve i as a factor.

To demonstrate that (11.27) is consistent with the elementary form of the uncertainty principle discussed earlier, set $\tilde{Q} = x$ and $\tilde{M} = \tilde{p} =$

$(\hbar/i)\, d/dx$. Then, for a test state ψ, we have

$$[\tilde{x}, \tilde{p}]\psi = x\frac{\hbar}{i}\frac{d\psi}{dx} - \frac{\hbar}{i}\frac{d}{dx}(x\psi)$$
$$= i\hbar\psi.$$

Hence,

$$\sigma_x^2\sigma_p^2 \geq \left(\frac{1}{2i}i\hbar\right)^2 = \left(\frac{\hbar}{2}\right)^2,$$

or

$$\sigma_x\sigma_p \geq \frac{\hbar}{2},$$

in agreement with the earlier observation.

In the case of a Hermitian (or self-adjoint) operator $\tilde{Q}$ with eigenvalue–eigenvector pairs $\{(\lambda_k, \varphi_k)\}_{k=1}^{\infty}$, we have seen that

$$\tilde{Q}\psi = \sum_{k=1}^{\infty}\lambda_k c_k\varphi_k, \qquad c_k = \langle\varphi_k, \psi\rangle.$$

The projection operators $\{P_k\}$ defined by

$$P_k\psi = \langle\varphi_k, \psi\rangle\,\varphi_k = c_k\varphi_k,$$

have the property that

$$\psi = I\psi = \sum_{k=1}^{\infty}c_k\varphi_k = \sum_{k=1}^{\infty}P_k\psi,$$

so that, symbolically,

$$\sum_{k=1}^{\infty}P_k = I = \text{ the identity.} \qquad (11.29)$$

For this reason, such a set of projections is called a *resolution of the identity*. The operators $\tilde{Q}$ can thus be represented as

$$\tilde{Q} = \sum_{k=1}^{\infty}\lambda_k P_k.$$

It is possible to develop an extension of these ideas to cases in which Q has a continuous spectrum.

11.7.1   Simultaneous Eigenfunctions

There is a special property of commuting operators called *the property of simultaneous eigenfunctions*. Suppose operators $\tilde{Q}$ and $\tilde{M}$ are such that ψ is an eigenfunction of both $\tilde{Q}$ and $\tilde{M}$; i.e. $\tilde{M}\psi = \lambda\psi$ and $\tilde{Q}\psi = \mu\psi$. Then

$$\tilde{Q}\tilde{M}\psi = \tilde{Q}\lambda\psi = \lambda\mu\psi = \tilde{M}\mu\psi = \tilde{M}\tilde{Q}\psi;$$

so $(\tilde{Q}\tilde{M} - \tilde{M}\tilde{Q})\psi = [\tilde{Q}, \tilde{M}]\psi = 0$. Thus, if operators have the same eigenfunctions, they commute.

The converse holds for compatible operators with nondegenerate eigenstates. If $\{\psi_n\}$ is a sequence of nondegenerate eigenstates of a Hermitian operator $\tilde{Q}$, which commutes with the Hermitian operator $\tilde{M}$, and $\tilde{Q}\psi_n = \lambda_n\psi_n$, then

$$\begin{aligned}
0 &= \big\langle \psi_m, [\tilde{M}, \tilde{Q}]\psi_n \big\rangle \\
&= \big\langle \psi_m, \tilde{M}\tilde{Q}\psi_n \big\rangle - \big\langle \psi_m, \tilde{Q}\tilde{M}\psi_n \big\rangle \\
&= \lambda_n \big\langle \psi_m, \tilde{M}\psi_n \big\rangle - \lambda_m \big\langle \psi_m, \tilde{M}\psi_n \big\rangle \\
&= (\lambda_n - \lambda_m)\big\langle \psi_m, \tilde{M}\psi_n \big\rangle.
\end{aligned}$$

Since the eigenstates are nondegenerate, for $n \neq m$ we have $\big\langle \psi_n, \tilde{M}\psi_m \big\rangle = 0$, which implies that $\big\langle \psi_n, \tilde{M}\psi_m \big\rangle \propto \delta_{nm}$, or that $\tilde{M}\psi_m$ is a constant times ψ_m. Thus, $\tilde{Q}$ and $\tilde{M}$ share eigenstates.

For the degenerate case, only a subset of the eigenstates of $\tilde{Q}$ may possibly be eigenstates of $\tilde{M}$.

CHAPTER 12

APPLICATIONS: THE HARMONIC OSCILLATOR AND THE HYDROGEN ATOM

12.1 Introductory Remarks

In this chapter, we consider applications of the theory discussed thus far to basic examples in quantum mechanics. These include the elementary example of the case of the harmonic oscillator for which the notion of ground states and quantization of energy levels is immediately derivable from Schrödinger's equation. Then the calculation of the quantum dynamics (the wave function) for the fundamental problem of the hydrogen atom is also described.

12.2 Ground States and Energy Quanta: The Harmonic Oscillator

Let us return to the simple case of a particle moving along a straight line with position coordinate $x = q$. We may then consider wave functions of the form

$$\Psi(q, t) = \psi(q)e^{iEt/\hbar}. \tag{12.1}$$

An Introduction to Mathematical Modeling: A Course in Mechanics, First Edition. By J. Tinsley Oden
© 2011 John Wiley & Sons, Inc. Published 2011 by John Wiley & Sons, Inc.

Then Schrödinger's equation reduces to

$$H\left(q, \frac{\hbar}{i}\frac{d}{dq}\right)\psi(q) = E\,\psi(q). \tag{12.2}$$

Thus, the energy E is an eigenvalue of the Hamiltonian H.

The Hamiltonian, we recall, is of the form

$$H(q,p) = \frac{p^2}{2m} + V(q)$$
$$= -\frac{\hbar^2}{2m}\frac{d^2}{dq^2} + V(q). \tag{12.3}$$

Hence, to complete the definition of the operator H, we need to specify the potential V. A classical and revealing example concerns the case in which V represents the potential energy of a harmonic oscillator vibrating with angular frequency ω:

$$V(q) = \frac{1}{2}m\omega^2 q^2. \tag{12.4}$$

Then (12.2) becomes

$$\left(-\frac{\hbar^2}{2m}\frac{d^2}{dq^2} + \frac{1}{2}m\omega^2 q^2\right)\psi = E\psi, \tag{12.5}$$

or

$$\left(\frac{d^2}{dq^2} + \lambda - \alpha^2 q^2\right)\psi = 0, \tag{12.6}$$

where

$$\lambda = 2mE/\hbar^2, \qquad \alpha = m\omega/\hbar. \tag{12.7}$$

The (normalizable) solutions of (12.6) are of the form

$$\psi(q) = a(q)e^{-\alpha q^2/2},$$

and, for the simplest case $a(q) = a_0$, a direct substitution reveals that the corresponding eigenvalue is

$$\lambda = \lambda_0 = \alpha = m\omega/\hbar.$$

Thus, the corresponding energy is

$$E = E_0 = \frac{1}{2}\hbar\omega. \tag{12.8}$$

This lowest possible energy corresponds to the *ground state* and is seen to be half of Planck's energy quantum.

The remaining solutions of the wave equation can be found by standard power-series expansions (method of Frobenius) to be of the form

$$\psi_n(q) = \left(\frac{m\omega}{\pi\hbar}\right)^{1/4} \frac{1}{\sqrt{n!2^n}} H_n(\xi)e^{-\xi^2/2}, \qquad n = 0, 1, 2, \ldots, \tag{12.9}$$

where $H_n(\cdot)$ are Hermite polynomials of order n and $\xi = q(m\omega/\hbar)^{1/2}$. The corresponding energy states are

$$\boxed{E_n = (n + \tfrac{1}{2})\hbar\omega.} \tag{12.10}$$

Thus, energy occurs in *quantized states* distinguished by different integer values of n, and the lowest energy is the ground state corresponding to $n = 0$. This elementary example again shows how the realization of discrete quantum states is a natural mathematical consequence of the discrete spectrum of this second-order, linear, elliptic differential operator.

12.3 The Hydrogen Atom

The harmonic oscillator provides an elementary example of how ground and quantized energy states are natural mathematical properties of eigenvalues of Schrödinger's equation, but the particular choice of the potential V was, to an extent, contrived. We now turn to one of the most important examples in quantum mechanics that provides the building blocks for quantum chemistry: Schrödinger's equation for the hydrogen atom. This involves a Hamiltonian with a potential that characterizes the energy of a charged particle, an electron, orbiting a nucleus with a charge of equal magnitude but opposite sign.

For the hydrogen atom, we have

$$H = -\frac{\hbar^2}{2m_p}\Delta_{\mathbf{r}_p} - \frac{\hbar^2}{2m_e}\Delta_{\mathbf{r}_e} + V(r),$$

where $(m_p, \mathbf{r}_p)$, $(m_e, \mathbf{r}_e)$ are the mass-position pairs of the proton and the electron of hydrogen with respect to a fixed point O, and

$$V(r) = -\frac{e^2}{4\pi\epsilon_0 r}, \qquad r = |\mathbf{r}_e - \mathbf{r}_p|. \tag{12.11}$$

If we relocate the origin at the center of mass $\mathbf{R} = (m_e\mathbf{r}_e + m_p\mathbf{r}_p)/(m_e + m_p)$, the Hamiltonian assumes the form

$$H = -\frac{\hbar^2}{2M}\Delta_{\mathbf{R}} - \frac{\hbar^2}{2\tilde{m}}\Delta_{\mathbf{r}} + V(r), \tag{12.12}$$

where $M = m_e + m_p$, $\tilde{m} = m_e m_p/M$, and $\mathbf{r} = \mathbf{r}_e - \mathbf{r}_p$. The mass of an electron is known to be $m_e = 9.10939 \times 10^{-31}$ kg, while that of a proton is $m_p = 1.67262 \times 10^{-27}$ kg. Since the proton mass is over 1800 times that of the electron, $\tilde{m} \approx m_e$ (in fact, $\tilde{m} = 1836m_e/1837 = 0.99946\, m_e$ [39, p. 88]), we ignore the motion of the proton and put the origin O precisely at the proton so that $\mathbf{r} = \mathbf{r}_e$. Then H reduces to

$$H = -\frac{\hbar^2}{2m}\Delta + V(r),$$

with Δ the three-dimensional Laplacian ($\Delta = \Delta_{\mathbf{r}}$), $m = m_e$, and $V = -e^2/4\pi\epsilon_0 r$. Schrödinger's equation for the single electron is then

$$i\hbar\frac{\partial\Psi}{\partial t} + \frac{\hbar^2}{2m}\Delta\Psi - V\Psi = 0. \tag{12.13}$$

Setting

$$\Psi(x, y, z, t) = \psi(x, y, z)e^{-iEt/\hbar}, \tag{12.14}$$

we have

$$-\frac{\hbar^2}{2m}\Delta\psi + V\psi = E\psi. \tag{12.15}$$

The general solution of (12.15) is, in principle, obtained by superposition: $\Psi = \sum_n c_n \Psi_n$ with c_n determined from initial data ($\Psi(x, y, z, 0)$) and $H\psi_n = E_n\psi_n$. But in this case, details of the properties of these eigenfunctions and eigenvalues play a fundamental role in shaping our understanding of the atomic structure of matter.

12.3.1 Schrödinger Equation in Spherical Coordinates

Because of spherical symmetry, it is convenient to transform the Schrödinger equation (12.15) into spherical coordinates (r, θ, ϕ), r being the radial distance from the nucleus at the origin, θ being the inclination angle with respect to the z axis, and ϕ the azimuthal angle from the x axis to the projection of the radius onto the xy plane. In these coordinates, the time-independent Schrödinger equation becomes (with ψ_n replaced by ψ for simplicity)

$$-\frac{\hbar^2}{2m}\left[\frac{1}{r^2}\frac{\partial}{\partial r}\left(r^2\frac{\partial\psi}{\partial r}\right) + \frac{1}{r^2 \sin\theta}\frac{\partial}{\partial\theta}\left(\sin\theta\frac{\partial\psi}{\partial\theta}\right) + \frac{1}{r^2 \sin^2\theta}\frac{\partial^2\psi}{\partial\phi^2}\right]$$
$$+ V\psi = E\psi. \tag{12.16}$$

We attempt to solve (12.16) by the method of separation of variables, wherein we separate the radial dependence from the (θ, ϕ)-dependence by assuming that ψ is of the form

$$\psi(r, \theta, \phi) = R(r)Y(\theta, \phi). \tag{12.17}$$

The standard process is to substitute (12.17) into (12.16), divide by RY, and then multiply by $-2mr^2/\hbar^2$ to obtain the sum,

$$\mathcal{A}(R) + \mathcal{B}(Y) = 0, \tag{12.18}$$

where $\mathcal{A}$ and $\mathcal{B}$ are the differential expressions,

$$\mathcal{A}(R) = R^{-1}\frac{d}{dr}\left(r^2\frac{dR}{dr}\right) - \frac{2mr^2}{\hbar^2}[V(r) - E], \tag{12.19}$$

$$\mathcal{B}(Y) = Y^{-1}\left[(\sin\theta)^{-1}\frac{\partial}{\partial\theta}\left(\sin\theta\frac{\partial Y}{\partial\theta}\right) + (\sin^2\theta)^{-1}\frac{\partial^2 Y}{\partial\phi^2}\right]. \tag{12.20}$$

The standard argument now employed is that a function $\mathcal{A}(R)$ of only r can equal a function $-\mathcal{B}(Y)$ of only θ and ϕ, only if they both are equal to a constant, say $\pm k$, which can be shown to be an integer. We therefore arrive at two separate equations,

$$\mathcal{A}(R) = k,$$
$$\mathcal{B}(Y) = -k. \tag{12.21}$$

The constant k is sometimes called the separation constant (cf., e.g., Griffiths [21, p. 124]).

Later, we will also represent Y as a product of a function of θ and a function of ϕ in the same spirit.

12.3.2   The Radial Equation

Since $R(r)$ must be finite at $r = 0$ and must vanish as $r \to \infty$, we can choose in the limit the radial function to be of the form $R = a_0 e^{-\alpha r}$, a_0 and α being constants. Introducing this into (12.21) gives

$$\left(\alpha^2 + \frac{2mE}{\hbar^2} - \frac{2\alpha}{r} + \frac{2me^2}{4\pi\varepsilon_0\hbar^2 r} - \frac{k}{r^2} \right) R(r) = 0.$$

Upon equating like powers of r, we conclude that this equation is satisfied for $R(r) \neq 0$ if $k = 0$ and

$$\alpha = \frac{me^2}{4\pi\varepsilon_0\hbar^2},$$

from which we also obtain

$$E = E_1 = -\frac{me^4}{(4\pi\varepsilon_0)^2 2\hbar^2}. \tag{12.22}$$

The energy E_1 in (12.22) is the *ground state energy* of the hydrogen atom.

The full form of the radial solution is obtained by considering trial solutions of the form $(c_1 r + c_2 r^2 + c_3 r^3 + \cdots)e^{-\alpha r}$ and terminating the expansion so that the function remains normalizable. In particular, we will normalize R in the sense that $\int_0^\infty |R|^2 r^2 \, dr = 1$. Replacing k by the integers $k = \ell(\ell + 1), \ell = 0, 1, \ldots, n - 1$, we can write the general solutions of $\mathcal{A}(R) = \ell(\ell + 1)$ as

$$
R_{n\ell}(r) = -\left[\frac{4(n - \ell - 1)!}{n^4 a_0^3 [(n + \ell)!]^3}\right]^{1/2} \left(\frac{2r}{na_0}\right)^\ell e^{-r/na_0} L_{n+\ell}^{1+2\ell}(2r/na_0),
$$

$$
n = 1, 2, \ldots, \quad \ell = 0, 1, \ldots, n - 1,
$$

(12.23)

where a_0 is the so-called *Bohr radius* (because it is the radius of the orbit of lowest energy in Bohr's 1913 model of the hydrogen atom),

$$
a_0 = \frac{4\pi\varepsilon_0 \hbar^2}{me^2}
$$

(12.24)

$(a_0 \approx 0.529\overset{\circ}{A})$, and $L_j^i(\cdot)$ are the *associated Laguerre polynomials*:

$$
L_{j-i}^i(x) = (-1)^i \left(\frac{d}{dx}\right)^i L_j(x),
$$

$$
L_q(x) = e^x \left(\frac{d}{dx}\right)^q (e^{-x} x^q).
$$

(12.25)

Examples are:

$$
L_0^0 = 1, \qquad L_0^2 = 2, \qquad L_1^0 = 1 - x,
$$
$$
L_1^2 = 18 - 6x, \quad L_2^0 = x^2 - 4x + 2, \quad \text{etc.}
$$

The range of the index ℓ terminates at $n - 1$ so that $R_{n\ell}$ is normalizable. [Note that in some treatments of this subject, a slightly different formula for $R_{n\ell}(r)$ is given in terms of generalized Laguerre polynomials which are defined differently than the associated polynomials used here.]

The indices n are called the *principal quantum numbers* and ℓ are called the *azimuthal quantum numbers*. The energy eigenvalues are independent of ℓ and are given by

$$
E_n = -\frac{1}{n^2}\left[\frac{me^4}{2\hbar^2(4\pi\varepsilon_0)^2}\right] = \frac{E_1}{n^2}, \qquad n = 1, 2, \ldots.
$$

(12.26)

The so-called *spectra* (or *line spectra*) of atoms describe the electromagnetic radiation that atoms emit when excited. For the hydrogen atom, the spectrum characterizes transitions between its possible energy states. The energy differences between states is released as a photon of energy $h\nu$ with wave number $\kappa = \nu/c$. The formula (12.26) is called the *Bohr formula*. From it, we can calculate the wavelength λ associated with transition between the quantum states n' and n''. According to Planck's formula, the energy of a photon is proportional to its frequency: $E = h\nu = ch/\lambda$. So

$$\frac{1}{\lambda} = R_0\left[\left(\frac{1}{n''}\right)^2 - \left(\frac{1}{n'}\right)^2\right], \tag{12.27}$$

where $R_0 = (m/4\pi c\hbar^3)(e^2/4\pi\varepsilon_0)^2 = 1.097 \times 10^7 m^{-1}$. The constant R_0 is known as *Rydberg's constant* and (12.27) is *Rydberg's formula* for the hydrogen spectrum. For the transition, $n_1 \leftarrow n_2$, the wave number of the emitted radiation is $R_0((1/n_1)^2 - (1/n_2)^2)$. For $n_1 = 1$ fixed, and $n_2 = 2, 3, \ldots$, we obtain the so-called Lyman series for ultraviolet light frequencies; for $n_1 = 2, n_2 = 3, \ldots$, the visible light Balmer series is obtained. Infrared begins with the Paschen series at $n_1 = 3, n_2 = 4, 5, \ldots$, and the Brackett series starts at $n_1 = 4, n_2 = 5, 6, \ldots$. When the electron is removed, the ionized state is obtained. The minimum energy required to ionize the atom is *ionization energy*, which is

$$E = hcR_0\left(1 - 1/n''^2\right)_{n'' \to \infty} = hcR_0.$$

12.3.3 The Angular Equation

We return now to the angular equation, $(12.21)_2$, with $k = \ell(1 + \ell)$. Again we employ separation of variables, setting

$$Y(\theta, \phi) = \Theta(\theta)\,\Phi(\phi).$$

We separate $\mathcal{B}(Y) = -\ell(\ell+1)$ into differential equations in θ and ϕ with separation constant m^2, m found to be an integer by standard arguments: $m = 0, \pm 1, \pm 2, \ldots$ (of course, m is not to be confused with the mass

$m = m_0$ of the electron). The integers m are called *magnetic quantum numbers*. The resulting equations are

$$\sin\theta \frac{d}{d\theta}\left(\sin\theta \frac{d\Theta}{d\theta}\right) + \left[\ell(1+\ell)\sin^2\theta - m^2\right]\Theta = 0, \qquad (12.28)$$

$$\frac{d^2\Phi}{d\phi^2} + m^2\Phi = 0. \qquad (12.29)$$

The solutions of these equations are

$$\begin{aligned}\Theta(\theta) &= \Theta_{\ell m}(\theta) = A\, P_\ell^m(\cos\theta), \\ \Phi(\phi) &= \Phi_m(\phi) = e^{im\phi},\end{aligned} \quad \text{for} \quad \begin{cases} \ell = 0, 1, \ldots, n-1, \\ m = 0, \pm 1, \pm 2, \ldots, \pm\ell, \end{cases}$$

where A is a constant and P_ℓ^m are the *associated Legendre polynomials*:

$$\begin{aligned} P_\ell^m(x) &= (1-x^2)^{|m|/2}\left(\frac{d}{dx}\right)^{|m|} P_\ell(x), \\ P_\ell(x) &= \frac{1}{2!\ell!}\left(\frac{d}{dx}\right)^\ell (x^2-1)^\ell, \\ x &\in [-1,1]. \end{aligned} \qquad (12.30)$$

Thus, the angular solutions are

$$\boxed{\begin{aligned} Y_{\ell m}(\theta,\phi) &= \Theta_{\ell m}(\theta)\,\Phi_m(\phi) = \delta_m \sqrt{\frac{(1+2\ell)(\ell-|m|)!}{4\pi(\ell+|m|)!}}\, e^{im\phi} P_\ell^m(\cos\theta) \\ \delta_m &= (-1)^m \text{ for } m \geq 0, \quad \delta_m = 1 \text{ for } m \leq 0, \\ \ell &= 0, 1, \ldots, n-1 \quad \text{and} \quad m = 0, \pm 1, \pm 2, \ldots, \pm\ell. \end{aligned}}$$

$$(12.31)$$

The functions $Y_{\ell m}$ are called *spherical harmonics*. They are mutually orthogonal in the sense that

$$\int_0^{2\pi}\int_0^\pi Y_{\ell m}(\theta,\phi)^* Y_{\ell' m'}(\theta,\phi)\sin\theta\, d\theta d\phi = \delta_{\ell\ell'}\,\delta_{mm'}. \qquad (12.32)$$

12.3.4   The Orbitals of the Hydrogen Atom

Summing up, the complete hydrogen wave functions are

$$
\boxed{
\begin{aligned}
\psi_{n\ell m} &= R_{n\ell}(r)\, Y_{\ell m}(\theta, \phi), \\
n &= 1, 2, \ldots, \\
\ell &= 0, 1, \ldots, n-1, \\
m &= 0, \pm 1, \pm 2, \ldots, \pm \ell,
\end{aligned}
}
\tag{12.33}
$$

where $R_{n\ell}(r)$ is given by (12.23) and $Y_{\ell m}(\theta, \phi)$ by (12.31). The corresponding wave functions are mutually orthonormal:

$$
\int_0^\infty \int_0^{2\pi} \int_0^\pi \psi_{n\ell m}^* \, \psi_{n'\ell'm'} \, r^2 dr \, \sin\theta \, d\theta \, d\phi = \delta_{nn'} \, \delta_{\ell\ell'} \, \delta_{mm'}. \tag{12.34}
$$

The functions $\psi_{n\ell m}$ are eigenfunctions of the Hamiltonian for the hydrogen atom (assuming $\tilde{m} \approx m_e = m$) and the eigenvalues correspond to energies of the orbital states. We refer to these eigenstates as *orbitals*.

12.3.5   Spectroscopic States

Eigenfunctions with degenerate eigenvalues correspond to cases in which two or more eigenfunctions coincide for the same eigenvalue. As a means for categorizing various orbitals and energy states and their energy levels, a universally used spectroscopic state convention is employed which assigns orbital types to various quantum numbers. The following observations and conventions apply to hydrogen orbitals:

1. Since

$$
1 = \langle \psi^*, \psi \rangle = \int_0^\infty \int_0^{2\pi} \int_0^\pi R^* Y^* R Y r^2 \, dr \, \sin\theta \, d\theta \, d\varphi
$$

if

$$
\int_0^{2\pi} \int_0^\pi Y^* Y \, \sin\theta \, d\theta \, d\varphi = 1,
$$

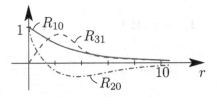

Figure 12.1: Components of the radial wave function, $R_{n\ell}(r)$, indicating rapid decay with increasing r.

we can define as the radial probability

$$\mathbb{P}_{n\ell}(r) = r^2 \int_0^{2\pi} \int_0^{\pi} |\psi_{n\ell m}|^2 \sin\theta \, d\theta \, d\phi = r^2 R_{n\ell}^2(r),$$

which decays rapidly as r increases. The portion of the domain in $\mathbb{R}^3$ of each orbital containing, say, 90% of the probability density inside the boundary, define standard geometric shapes of the orbitals for various energy levels (see Figure 12.1).

2. For $\ell = 0$, we obtain the s orbitals, $1s, 2s, \ldots, ns$, for various values of the principal quantum number n. These are spherical domains (under the conditions of the convention 1 above), with $1s$ corresponding to the (lowest) ground state energy. Only the s-orbitals are nonzero at $r = 0$. The ground state energy E_1 is negative, so energies increase with principal quantum number n as indicated in Table 12.1.

3. Orbitals for the same value of n but different ℓ and m are degenerate. For $\ell \neq 0$, $m = 0$, the so-called z orbitals are obtained (denoted $2p_z, 3d_z, \ldots$). The $2p$ orbitals ($n = 2, \ell = 1$), for example, comprise one real and two complex-valued functions ($m = 0$ and $m = \pm 1$):

$$
\begin{aligned}
2p(-1): \quad & \sqrt{3/8\pi}\, R_{21}(r) \sin\theta e^{-i\phi}, \\
2p(0): \quad & \sqrt{3/4\pi}\, R_{21}(r) \cos\theta, \\
2p(+1): \quad & -\sqrt{3/8\pi}\, R_{21}(r) \sin\theta e^{i\phi}.
\end{aligned}
$$

Table 12.1: Hydrogen orbitals.

Quantum number		Orbital name	Degeneracy	Energy
n	ℓ			(Eq. (12.26))
1	0	$1s$	1	E_1
2	0	$2s$	1	$E_1/4$
2	1	$2p$	3	$E_1/4$
3	0	$3s$	1	$E_1/9$
3	1	$3p$	3	$E_1/9$
3	2	$3d$	5	$E_1/9$
$\vdots$	$\vdots$	$\vdots$	$\vdots$	$\vdots$

The $2p(0)$ corresponds to the $2p_z$ orbital, and linear combinations of $2p(+1)$ and $2p(-1)$ correspond to the $2p_x$ and $2p_y$ orbitals:

$$2p_x = -\frac{1}{\sqrt{2}} \left[2p(+1) - 2p(-1) \right]: \quad \sqrt{3/4\pi}\, R_{21}(r) \sin\theta \cos\phi,$$

$$2p_y = -\frac{1}{i\sqrt{2}} \left[2p(+1) + 2p(-1) \right]: \quad \sqrt{3/4\pi}\, R_{21}(r) \sin\theta \sin\phi.$$

Similar linear combinations of other degenerate pairs lead to the $3p, 3d, 4f, \ldots$ orbitals. The literature in quantum chemistry suggests that s, p, d, f stand for "sharp," "principal," "diffuse," and "fundamental," respectively. We will take the subject up in more detail in Chapter 14 in our study of atomic structure and the periodic table.

Examples of some of the hydrogen wave functions are given in Table 12.2.

Table 12.2: Wave function components. In the formulas, we have $\gamma = \pi^{-1/2}(Z/a_0)^{3/2}$ and $\zeta = Zr/a_0$, where eZ is the charge at origin (in our case $Z = 1$) and $a_0 = 4\pi\epsilon_0\hbar^2/me^2$ is the Bohr radius.

| Quantum number | | | Orbital name | Wave function component |
n	ℓ	m		$\psi_{n\ell m}$
1	0	0	$1s$	$\gamma e^{-\zeta}$
2	0	0	$2s$	$\frac{1}{2\sqrt{2}}\gamma(1 - \zeta/2)e^{-\zeta/2}$
2	1	0	$2p_z$	$\frac{1}{4\sqrt{2}}\gamma\zeta e^{-\zeta/2}\cos\theta$
2	1	± 1	$2p_{\pm 1}$	$\mp\frac{1}{8}\gamma\zeta e^{-\zeta/2}e^{\pm i\phi}\sin\theta$
3	0	0	$3s$	$\frac{1}{81\sqrt{3}}\gamma(27 - 18\zeta + 2\zeta^2)e^{-\zeta/3}$
3	1	0	$3p_z$	$\frac{2\sqrt{2}}{81}\gamma(6\zeta - \zeta^2)e^{-\zeta/3}\cos\theta$
$\vdots$	$\vdots$	$\vdots$	$\vdots$	$\vdots$

CHAPTER 13

SPIN AND PAULI'S PRINCIPLE

13.1 Angular Momentum and Spin

Up to this point, the *linear* momentum **p** of a system has played the role of a fundamental dynamical property of quantum systems. From classical mechanics, however, we know that other types of momenta can exist by virtue of the moment of momentum vectors relative to a fixed point in space and by virtue of their spin about a trajectory of the particle. We shall now examine how these properties manifest themselves in quantum systems.

We know from classical dynamics of rigid bodies that when a body (such as a satellite) moves along a trajectory in three-dimensional space, it has angular momentum by virtue of two effects: the moment **L** of the momentum vector tangent to its trajectory about a fixed origin (such as the earth's center) and the angular momentum **S** due to spin of the body about its axis. The former is called the *extrinsic angular momentum* and is given by

$$\mathbf{L} = \mathbf{q} \times \mathbf{p}, \tag{13.1}$$

where **q** is the position of the mass center of the body relative to the origin and **p** is its linear momentum, and the latter is called the *intrinsic*

An Introduction to Mathematical Modeling: A Course in Mechanics, First Edition. By J. Tinsley Oden
© 2011 John Wiley & Sons, Inc. Published 2011 by John Wiley & Sons, Inc.

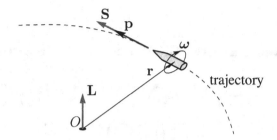

Figure 13.1: Extrinsic angular momentum **L** and intrinsic angular momentum **S** of a rigid body.

angular momentum and is given by

$$\mathbf{S} = I\omega, \tag{13.2}$$

where I is the moment of inertia of the body about its (longitudinal) spin axis and ω is its angular velocity. These are illustrated in Fig. 13.1. Note that **S** can be directed forward toward the direction of motion if the body spins clockwise or backward if it spins counterclockwise.

Analogously, it so happens that an electron moving in an orbital trajectory also has two types of angular momenta: (1) the extrinsic momentum, $\mathbf{L} = \mathbf{q} \times \mathbf{p}$, where $\mathbf{q} = \mathbf{r}$ is now the distance from a fixed origin, generally the center of the nucleus, and **p** is the electron momentum ($\mathbf{p} = (\hbar/i)\partial/\partial q$), and (2) an intrinsic momentum **S**. While **L** is obviously a dynamical variable of a quantum system, so we can use the ideas developed up to now to study its properties, the existence of the spin **S** is established in relativistic quantum mechanics and falls outside the scope of the non-relativistic theory we study here. We introduce **S** within our theory as a postulate:

*Every elementary particle in a quantum system has an intrinsic spin **S**; in the case of an electron, the spin has magnitude $\hbar/2$.*

We provide more details on this specific type of intrinsic spin in Section 13.3. The general approach is to derive dynamical properties of the extrinsic angular momentum **L** and to postulate that **S** shares exactly the same properties up to a point; the spectral properties (i.e., the structure of the eigenvectors and eigenvalues) of **L** and **S** are quite different. Indeed, the eigenfunctions of **L** are functions we recognize from our study of the hydrogen atom while those of **S** are multiples of pairs of integers.

13.2 Extrinsic Angular Momentum

We begin with deriving properties of the extrinsic angular momentum, and will then assign analogous properties to the spin. Thus,

$$\mathbf{L} = \mathbf{q} \times \mathbf{p},$$

and in Cartesian coordinates we have

$$L_j = \varepsilon_{rsj} q_r \frac{\hbar}{i} \frac{\partial}{\partial q_s}, \qquad (13.3)$$

ε_{rsj} being the permutation symbol and we sum on repeated indices. Thus

$$L_1 = \frac{\hbar}{i} \left(q_2 \frac{\partial}{\partial q_3} - q_3 \frac{\partial}{\partial q_2} \right),$$

$$L_2 = \frac{\hbar}{i} \left(q_3 \frac{\partial}{\partial q_1} - q_1 \frac{\partial}{\partial q_3} \right), \qquad (13.4)$$

$$L_3 = \frac{\hbar}{i} \left(q_1 \frac{\partial}{\partial q_2} - q_2 \frac{\partial}{\partial q_1} \right).$$

Since we will want to consider the motion of particles in spherical orbits, we can also write these operators in spherical coordinates as follows:

$$L_1 = \frac{\hbar}{i} \left(-\sin\phi \frac{\partial}{\partial\theta} - \cos\phi \cot\theta \frac{\partial}{\partial\phi} \right),$$

$$L_2 = \frac{\hbar}{i} \left(\cos\phi \frac{\partial}{\partial\theta} - \sin\phi \cot\theta \frac{\partial}{\partial\phi} \right), \qquad (13.5)$$

$$L_3 = \frac{\hbar}{i} \frac{\partial}{\partial\phi}.$$

As a first set of properties, we begin with commutativity.

Theorem 13.A

1. The extrinsic angular momentum operators L_j of (13.3) do not commute, and indeed

$$[L_1, L_2] = i\hbar L_3, \quad [L_2, L_3] = i\hbar L_1, \quad [L_3, L_1] = i\hbar L_2. \quad (13.6)$$

2. The operators L_j do commute with the operator

$$L^2 = L_1^2 + L_2^2 + L_3^2. \quad (13.7)$$

Proof 1. For a given test function φ, a direct calculation gives

$$
\begin{aligned}
[L_1, L_2]\varphi &= \left(\frac{\hbar}{i}\right)^2 \left\{ \left(q_2\frac{\partial}{\partial q_3} - q_3\frac{\partial}{\partial q_2} \right)\left(q_3\frac{\partial}{\partial q_1} - q_1\frac{\partial}{\partial q_3} \right)\varphi \right.\\
&\quad \left. - \left(q_3\frac{\partial}{\partial q_1} - q_1\frac{\partial}{\partial q_3} \right)\left(q_2\frac{\partial}{\partial q_3} - q_3\frac{\partial}{\partial q_2} \right)\varphi \right\}\\
&= \left(\frac{\hbar}{i}\right)^2 (q_2\varphi_{,1} + q_2 q_3\varphi_{,31} - q_2 q_1\varphi_{,33} - q_3^2\varphi_{,12} + q_3 q_1\varphi_{,23}\\
&\quad - q_3 q_2\varphi_{,13} + q_3^2\varphi_{,21} + q_1 q_2\varphi_{,33} - q_1\varphi_{,2} - q_1 q_3\varphi_{,32})\\
&= \left(\frac{\hbar}{i}\right)^2 (q_2\varphi_{,1} - q_1\varphi_{,2})\\
&= i\hbar L_3\varphi,
\end{aligned}
$$

where we have used the shorthand notation, $\varphi_{,i} = \partial\varphi/\partial q_i$, $\varphi_{,ij} = \partial^2\varphi/\partial q_i\partial q_j$ for $i, j = 1, 2, 3$. Thus,

$$[L_1, L_2] = i\hbar L_3.$$

The remaining relations in (13.6) follow from a permutation of the indices.

2. That L^2 commutes with the L_j follows from the fact

$$
\begin{aligned}
[L^2, L_1] &= [L_1^2, L_1] + [L_2^2, L_1] + [L_3^2, L_1]\\
&= L_2[L_2, L_1] + [L_2, L_1]L_2 + L_3[L_3, L_1] + [L_3, L_1]L_3,
\end{aligned}
$$

because $[L_1^2, L_1] = 0$, $[L_2^2, L_1] = L_2^2 L_1 - L_1 L_2^2 = L_2[L_2, L_1] - [L_1, L_2]L_2$, etc. So

$$[L^2, L_1] = L_2(-i\hbar L_3) + (-i\hbar L_3)L_2 + L_3(i\hbar L_2) + (i\hbar L_2)L_3$$
$$= 0,$$

with analogous results for $[L^2, L_2]$ and $[L^2, L_3]$.   □

We know from Chapter 11 that the fact that the L_j do not commute, identifies them as incompatible observables and means that their variances satisfy the uncertainty inequality of the form,

$$\sigma_{L_1}^2 \sigma_{L_2}^2 \geq \left(\frac{1}{2i} \langle i\hbar L_3 \rangle \right)^2 = \frac{\hbar^2}{4} \langle L_3 \rangle^2,$$

or

$$\sigma_{L_1} \sigma_{L_2} \geq \frac{\hbar}{2} \langle L_3 \rangle. \tag{13.8}$$

The fact that L^2 and L_j do commute means that they share simultaneous eigenfunctions; i.e., there exists a φ and eigenvalues λ and μ such that

$$L^2 \varphi = \lambda \varphi \quad \text{and} \quad L_j \varphi = \mu \varphi \quad (j = 1, 2, 3).$$

This relation holds for one choice of index j; however, according to (13.6), the L_j does not commute with L_k for $j \neq k$ and L_j does not share eigenfunctions with L_k.

13.2.1 The Ladder Property: Raising and Lowering States

The operators L_1 and L_2 can be used to construct a family of angular momentum states that decrease or increase the eigenvalues of the operator L_3 by composition with the operators $L_\pm$ defined by

$$L_\pm = L_1 \pm iL_2. \tag{13.9}$$

L_+ and L_- commute with L^2 but not L_3. However, for any function φ which is an eigenfunction of L^2, and $L_3 \varphi = \mu \varphi$, then $L_\pm \varphi$ is an eigenfunction of L_3 and L^2, and after some algebraic manipulations using (13.4) (see, e.g., Griffith [21, pp. 147, 148] for complete details), we arrive at

$$L_3(L_\pm \varphi) = (\mu \pm \hbar) L_\pm \varphi. \tag{13.10}$$

Thus a family of eigenfunctions of L^2 can be generated by L_+, with L_+ raising the eigenvalue by $\hbar$ and L_- lowering it by $\hbar$. For a given eigenvalue λ of L^2, we obtain a *ladder of states* with each state separated by the distance $\hbar$. If we assign an index m to count the states corresponding to the eigenvalues of L_3, and ℓ for these of L^2, one can show that the eigenvalues of L^2 depend on $\ell(\ell+1)$ and those of L_3 depend on m.

What are the eigenfunctions of L^2 and L_3? Since they arise from angular momentum, let us return to the Schrödinger equation for a single electron given in (12.16) and consider the case of purely rotational motion represented by the case in which $r = r_0 = $ constant. The equation collapses to

$$-\frac{\hbar^2}{2I}\left\{(\sin \theta)^{-1}\frac{\partial}{\partial \theta}\left(\sin \theta \frac{\partial}{\partial \theta}\right) + (\sin^2 \theta)^{-1}\frac{\partial^2}{\partial \phi^2}\right\}\psi(\theta, \phi) = E\psi(\theta, \phi),$$
$$\tag{13.11}$$

where $I = mr_0^2$ is the rotary inertia. We know from the analysis of the hydrogen atom that the angular eigenfunctions are the spherical harmonics $Y_{\ell m}$ of (12.31). Introducing (13.5), we have

$$L^2 = L_1^2 + L_2^2 + L_3^2 = -\hbar^2\left\{(\sin \theta)^{-1}\frac{\partial}{\partial \theta}\left(\sin \theta \frac{\partial}{\partial \theta}\right) + (\sin^2 \theta)^{-1}\frac{\partial^2}{\partial \phi^2}\right\}.$$
$$\tag{13.12}$$

Thus, the rigid-rotor part of the Hamiltonian is $L^2/2I$. It follows that

$$L^2 Y_{\ell m} = \ell(1 + \ell)\hbar^2 Y_{\ell m}, \tag{13.13}$$

and, simultaneously,

$$L_3 Y_{\ell m} = \hbar m Y_{\ell m}. \tag{13.14}$$

According to (13.10), the raising and lowering operators $L_\pm$ produce a ladder of states, each separated from its neighbor by $\hbar$ in eigenvalues of L_3. We climb up the ladder by applying L_+ and down by applying L_-. There exist functions ψ_{top} for the top rung and ψ_{bot} for the bottom rung such that $L_+\psi_{\text{top}} = 0$ and $L_-\psi_{\text{bot}} = 0$. One can in fact show that $L^2\psi_{\text{top}} = \hbar^2\ell(1 + \ell)\psi_{\text{top}}$ and $L^2\psi_{\text{bot}} = \hbar^2\hat{\ell}(1 + \hat{\ell})\psi_{\text{bot}}$. There is an upper bound for the quantum number ℓ and a lower bound for the quantum number $\hat{\ell}$, separated by N integer steps; $\hat{\ell} = -\ell$. The eigenvalues of L_3 are $m\hbar$ and m ranges from $-\ell$ to ℓ, so at the Nth step, $\ell = -\ell + N$. In other words, $\ell_{\max} = N/2$, $N = 0, 1, 2, \ldots$. Thus, unlike the azimuthal indices ℓ for the hydrogen atom, ℓ now takes on integer and half-integer values, and m ranges from $-\ell$ to $+\ell$:

$$\ell = 0, \tfrac{1}{2}, 1, \tfrac{3}{2}, \ldots, \qquad m = -\ell, -\ell + 1, \ldots, \ell - 1, \ell. \qquad (13.15)$$

Since for angular momentum, the eigenfunctions are spherical harmonics, half integers do not appear, and we have the following theorem

Theorem 13.B *The eigenfunctions of the angular momentum operators L^2 and L_3 are the spherical harmonics $Y_{\ell m}$ of (12.31) and the eigenvalues are $\hbar^2\ell(1+\ell)$ and $\hbar m$, respectively, where $\ell = 0, 1, 2, \ldots$ and $m = -\ell, -\ell + 1, \ldots, \ell - 1, \ell$.*

13.3 Spin

Turning now to the spin **S** of an elementary particle, we postulate that its corresponding operator components are endowed with exactly the same commutator properties as the extrinsic angular momentum **L**. Thus:

*The spin **S** of every elementary particle in a quantum system has components that satisfy*

$$[S_1, S_2] = i\hbar S_3, \quad [S_2, S_3] = i\hbar S_1, \quad [S_3, S_1] = i\hbar S_2. \qquad (13.16)$$

Thus, the S_j do not commute, but as in the case of the extrinsic angular momentum operators L_j, S^2 commutes with S_3 and has common eigenvectors:

$$S^2 q_{sm} = \hbar^2 s(1+s) q_{sm}, \qquad S_3 q_{sm} = \hbar m q_{sm}. \qquad (13.17)$$

One can also show that

$$S_{\pm} q_{sm} = \hbar \left[s(1+s) - m(m+1) \right]^{1/2} q_{s,m+1}, \qquad (13.18)$$

where $S_{\pm} = S_1 \pm i S_2$, so the spin operators also have the property of raising and lowering states analogous to $L_{\pm}$. But the eigenvectors q_{sm} are not functions of (θ, ϕ); rather, they are quantities categorized by the half-integers values of indices s and m:

$$s = 0, \frac{1}{2}, 1, \frac{3}{2}, \ldots, \qquad m = -s, -s+1, \ldots, s.$$

Every elementary particle has a specific value s of spin, depending only on the type of particle: for pi mesons, $s = 0$; for electrons, $s = \frac{1}{2}$; for photons, $s = 1$; for delta baryons, $s = \frac{3}{2}$; for gravitons, $s = 2$; etc. But the spin value s is fixed for any particular particle type.

It follows that the cases of interest here are those for which $s = 1/2$, which is the spin of particles that make up ordinary matter: electrons, protons, neutrons. In this case, there are just two eigenstates corresponding to $s = 1/2$, $m = \pm 1/2$: $q_{1/2,1/2}$ called *spin up* and symbolically represented by $\uparrow$, and $q_{1/2,-1/2}$, called *spin down* and represented by $\downarrow$. The spin components are not viewed as scalar components of a vector, but now 2×2 matrices $\mathbf{S}_i$. Using these as basis vectors, the state of an $s = 1/2$ spin particle is the linear combination

$$\mathbf{X} = a\mathbf{X}_+ + b\mathbf{X}_-, \qquad \mathbf{X}_+ = \begin{bmatrix} 1 \\ 0 \end{bmatrix}, \qquad \mathbf{X}_- = \begin{bmatrix} 0 \\ 1 \end{bmatrix}. \qquad (13.19)$$

We observe that the corresponding eigenvalues of S_3 are $+\hbar/2$ for spin up and $-\hbar/2$ for spin down.

According to (13.17), we have

$$\mathbf{S}^2 \mathbf{X}_+ = \frac{3\hbar^2}{4} \mathbf{X}_+, \qquad \mathbf{S}^2 \mathbf{X}_- = \frac{3\hbar^2}{4} \mathbf{X}_-,$$

$$\mathbf{S}_3 \mathbf{X}_+ = \frac{\hbar}{2} \mathbf{X}_+, \qquad \mathbf{S}_3 \mathbf{X}_- = -\frac{\hbar}{2} \mathbf{X}_-. \qquad (13.20)$$

If $\mathbf{X} = a\mathbf{X}_+ + b\mathbf{X}_-$ corresponds to a general state of a particle and we attempt to measure, say, S_3, then we get a value of $\hbar/2$ with probability $|a|^2$ or $-\hbar/2$ with probability $|b|^2$. We must, therefore, have $|a|^2 + |b|^2 = 1$. The spins S_1, S_2 do not commute, so an attempt at a measurement of them yields either $\hbar/2$ or $-\hbar/2$ with equal probability.

The spin operators in (13.16) and (13.17) can now assume the form

$$\mathbf{S}_1 = \frac{\hbar}{2}\sigma_1, \quad \mathbf{S}_2 = \frac{\hbar}{2}\sigma_2, \quad \mathbf{S}_3 = \frac{\hbar}{2}\sigma_3, \tag{13.21}$$

where σ_i are *Pauli's spin matrices*:

$$\sigma_1 = \begin{bmatrix} 0 & 1 \\ 1 & 0 \end{bmatrix}, \quad \sigma_2 = \begin{bmatrix} 0 & -i \\ i & 0 \end{bmatrix}, \quad \sigma_3 = \begin{bmatrix} 1 & 0 \\ 0 & -1 \end{bmatrix}. \tag{13.22}$$

If we set $\mathbf{S}_\pm = \mathbf{S}_1 \pm i\mathbf{S}_2$, then $\mathbf{S}_1 = (\mathbf{S}_+ + \mathbf{S}_-)/2$, $\mathbf{S}_2 = (\mathbf{S}_+ - \mathbf{S}_-)/2i$, and the eigenvectors $\mathbf{X}_\pm$ of (13.19), called *eigenspins* or *spinors*, satisfy

$$\mathbf{S}_1\mathbf{X}_+ = \frac{\hbar}{2}\mathbf{X}_-, \qquad \mathbf{S}_1\mathbf{X}_- = \frac{\hbar}{2}\mathbf{X}_+, \\ \mathbf{S}_2\mathbf{X}_+ = -\frac{\hbar}{2i}\mathbf{X}_-, \qquad \mathbf{S}_2\mathbf{X}_- = \frac{\hbar}{2i}\mathbf{X}_+. \tag{13.23}$$

Thus,

$$\mathbf{S}^2 = \mathbf{S}_1^2 + \mathbf{S}_2^2 + \mathbf{S}_3^2 = \frac{3}{4}\hbar^2 \begin{bmatrix} 1 & 0 \\ 0 & 1 \end{bmatrix},$$

$$\mathbf{S}_3 = \frac{\hbar}{2}\sigma_3,$$

and the eigenspinors $\mathbf{X}_\pm$ are clearly eigenvectors of $\mathbf{S}_3$ with eigenvalues $\pm\hbar/2$.

The question that arises at this point is: How should electron spin be incorporated into solutions of Schrödinger's equation? This is typically accomplished by writing each one-electron wave function as the product of (a) spatial orbitals of the type discussed earlier and (b) a spin function that depends on the spin of the electron. The results are the so-called spin orbitals. So, for instance, if $\psi_{n\ell m}$ is a hydrogen quantum state and X_+ is a spin state, $\psi_{n\ell m}X_+$ is a spin orbital. The wave function for the complete quantum state of position and spin for n electrons is of the form

$$\psi(\mathbf{r}_1, \mathbf{r}_2, \ldots, \mathbf{r}_n)\, \mathbf{X}(\mathbf{s}_1, \mathbf{s}_2, \ldots, \mathbf{s}_n),$$

where $\mathbf{X}$ is the spinor that describes the spin state. This entire composite function must be antisymmetric with respect to interchange of two electrons.

Another common spin notation recognizes that the electron spin functions have a value of 0 or 1, depending on the quantum number m and are labeled α and β, with

$$\alpha(m) = \begin{cases} 1, & m = 1/2, \\ 0, & m = -1/2 \end{cases} \quad \text{and} \quad \beta(m) = \begin{cases} 0, & m = 1/2, \\ 1, & m = -1/2. \end{cases}$$

$$\tag{13.24}$$

In general, the angular spin momentum in an orbital is balanced in the sense that the spin-up momentum is balanced with a spin down. The result is that, in general:

Each spatial orbital can accommodate only (up to) two electrons with paired spins.

This principle provides a fundamental rule for describing the electronic structure of polyelectron atoms. We return to this structure later.

As a final remark in this section, we recall from Chapter 9 that a spinning charged elementary particle creates a magnetic dipole with a *magnetic dipole moment* $\boldsymbol{\mu}$ proportional to the angular spin momentum $\mathbf{S}$ (recall (9.13)). When placed in a magnetic field $\mathbf{B}$, the particle experiences a torque $\boldsymbol{\mu} \times \mathbf{B}$ with energy $-\boldsymbol{\mu} \cdot \mathbf{B}$. The constant $m_0 \ (= -e/m)$ of proportionality, $\boldsymbol{\mu} = m_0 \mathbf{S}$, is called the *gyromagnetic ratio*, and the Hamiltonian for a spinning particle at rest is $H = -m_0 \mathbf{B} \cdot \mathbf{S}$.

13.4 Identical Particles and Pauli's Principle

When we extend the ideas covered thus far to multielectron systems, we immediately encounter a question that has profound implications on the form and properties of the wave function. Consider a system of only two perfectly identical particles, particle 1 with coordinates r_1 and particle 2 with coordinates r_2. Ignoring for the moment spin, the wave function of this system is of the form

$$\Psi = \Psi(\mathbf{r}_1, \mathbf{r}_2, t),$$

with Hamiltonian,

$$H = -\frac{\hbar^2}{2m}\Delta_1 - \frac{\hbar^2}{2m}\Delta_2 + V(\mathbf{r}_1, \mathbf{r}_2). \qquad (13.25)$$

If $\Psi = \psi(\mathbf{r}_1, \mathbf{r}_2)e^{-iEt/\hbar}$, the spatial wave function ψ satisfies the time-independent Schrödinger equation,

$$H\psi = E\psi.$$

If particle 1 is in a state $\psi_1(\mathbf{r}_1)$ and particle 2 is in a state $\psi_2(\mathbf{r}_2)$, then

$$\psi(\mathbf{r}_1, \mathbf{r}_2) = \psi_1(\mathbf{r}_1)\psi_2(\mathbf{r}_2). \qquad (13.26)$$

But if we interchange the two particles (place 2 at location $\mathbf{r}_1$ and 1 at location $\mathbf{r}_2$), we must arrive at the same dynamical system. We must, in other words, describe the wave function in a way that is noncommittal to which particle is in which state. Either of the following two combinations fulfill the requirement:

$$\psi_{\pm}(\mathbf{r}_1, \mathbf{r}_2) = C\Big(\psi_1(\mathbf{r}_1)\psi_2(\mathbf{r}_2) \pm \psi_2(\mathbf{r}_1)\psi_1(\mathbf{r}_2)\Big). \qquad (13.27)$$

Thus, in this example, quantum mechanics admits two kinds of wave functions for systems of two identical particles, one for which the sign in (13.27) is positive and one in which the sign is negative. Then,

$$\begin{aligned} \psi_+(\mathbf{r}_1, \mathbf{r}_2) &= +\psi_+(\mathbf{r}_2, \mathbf{r}_1), \\ \psi_-(\mathbf{r}_1, \mathbf{r}_2) &= -\psi_-(\mathbf{r}_2, \mathbf{r}_1). \end{aligned} \qquad (13.28)$$

It turns out that experimental observations reveal that the positive case ψ_+ corresponds to particles which have integer spins (which we call *bosons*) and the negative case ψ_- corresponds to half-integer spins, such as electrons (which we call *fermions*).

A more concise way to differentiate between the two possibilities is to introduce the *permutation operator*

$$P_{12}\psi(\mathbf{r}_1, \mathbf{r}_2) = \psi(\mathbf{r}_2, \mathbf{r}_1). \tag{13.29}$$

This operator has the following fundamental properties:

1. $P_{12} = P_{12}^2$ (P_{12} is a projection).

2. The eigenvalues of P_{12} are ± 1.

3. P_{12} commutes with the Hamiltonian H:

$$[H, P_{12}] = 0. \tag{13.30}$$

Thus, P and H are compatible observables and we can find a set of functions that are simultaneous eigenstates of both operators.

The wave functions can therefore be either symmetric or antisymmetric with respect to permutation of any two identical particles.

The observables $|\psi|^2$ and $|P_{12}\psi|^2 = |\psi|^2$ are the same physically possible quantities in either case.

These considerations and experimental evidence lead us to:

Pauli's Principle *Every elementary particle has an intrinsic angular momentum called its spin. Those with integer spins are called bosons, and the corresponding wave function is symmetric with respect to permutations of identical particles* $(P_{12}\psi = \psi)$. *Those with half-integer spins are called fermions, and the corresponding wave function is antisymmetric with respect to permutations of identical particles* $(P_{12}\psi = -\psi)$.

Returning to (13.27), suppose we have two electrons, 1 and 2, with $\psi_1 = \psi_2$. Then if one has two identical fermions, $\psi_-(\mathbf{r}_1, \mathbf{r}_2) = C\big(\psi_1(\mathbf{r}_1)\psi_1(\mathbf{r}_2) - \psi_1(\mathbf{r}_1)\psi_1(\mathbf{r}_2)\big) = 0$, which is a contradiction. Thus,

Two identical fermions cannot occupy the same state.

This observation is referred to as *Pauli's Exclusion Principle*.

We recall that pi mesons and photons are examples of elementary particles with integer spins, while electrons and protons have half-integer spins. Thus, electrons and protons, on which we focus this discussion, are fermions and have antisymmetric wave functions under permutations of identical particles. In the case of n identical particles, we can introduce the permutation operators,

$$P_{i_1 i_2 \cdots i_n} \psi(\mathbf{r}_1, \mathbf{r}_2, \ldots, \mathbf{r}_n) = \varepsilon_{i_1 i_2 \cdots i_n} \psi(\mathbf{r}_{i_1}, \mathbf{r}_{i_2}, \ldots, \mathbf{r}_{i_n})$$

(no sum on repeated indices), (13.31)

where $i_1 i_2 \cdots i_n$ is a permutation of the integers $1, 2, \ldots, n$ and $\varepsilon_{i_1 i_2 \cdots i_n}$ is the permutation symbol equal to $+1$ or -1 if, respectively, $i_1, i_2, \ldots, i_n$ is an even (symmetric) or odd (antisymmetric) permutation of $123 \cdots n$, and zero otherwise.

13.5   The Helium Atom

The next level of complexity beyond the hydrogen atom is the helium atom, which consists of two electrons, labeled 1 and 2, with positions r_1 and r_2 in space, relative to the origin at the nucleus. The Hamiltonian is

$$H = -\frac{\hbar^2}{2m}\Delta_1 - \frac{\hbar^2}{2m}\Delta_2 - \frac{Ze^2}{4\pi\varepsilon_0 r_1} - \frac{Ze^2}{4\pi\varepsilon_0 r_2} + \frac{e^2}{4\pi\varepsilon_0 r_{12}}, \qquad (13.32)$$

where Ze is the charge of the nucleus and $r_{12} = |r_2 - r_1|$. Schrödinger's equation,

$$H\psi = E\psi,$$

is not separable due to the presence of the electron–electron repulsive term involving $1/r_{12}$.

Unfortunately, for this system and all others excluding the hydrogen atom, no exact solutions exist. We must therefore resort to approximations. Among the most powerful and successful approximation method is the Rayleigh–Ritz method, known in quantum mechanics literature as simply "the variational method." We give a brief account of this method in the next section. The idea is simple: Minimize the energy E over a class of *admissible functions* in which the wave function belongs. It is this latter consideration, the identification of admissible functions, that now presents a formidable problem, since admissible functions must be possible spin orbitals or combinations of spin orbitals that are $L^2(\mathbb{R}^3)$ and, importantly, satisfy the antisymmetric requirement of fermions. The process of seeking such admissible functions gives us license to explore various approximations of Schrödinger's equation for the helium atom.

One common approximation of (13.32) that leads to an illustration of how to cope with the antisymmetric issue results when we ignore the $1/r_{12}$ term and consider the separable Hamiltonian,

$$(H_1 + H_2)\,\psi(\mathbf{r}_1, \mathbf{r}_2) = E\psi(\mathbf{r}_1, \mathbf{r}_2),$$

where

$$H_i = -\frac{\hbar^2}{2m}\Delta_i - \frac{Ze^2}{4\pi\varepsilon_0 r_i}, \qquad i = 1, 2.$$

For this separable approximation, the wave function can be written as the product of the individual one-electron hydrogen wave functions; symbolically,

$$\psi(\mathbf{r}_1, \mathbf{r}_2) \cong \psi_{n\ell m}(\mathbf{r}_1)\, \psi_{n'\ell'm'}(\mathbf{r}_2)$$

for various values of n, n', ℓ, ℓ', m, and m'.

Let us consider as a first approximation the case in which each factor is just the $1s$ orbital,

$$\varphi_{1s}(\mathbf{r}_1) = \psi_{100}(\mathbf{r}_1) \quad \text{and} \quad \varphi_{1s}(\mathbf{r}_2) = \psi_{100}(\mathbf{r}_2).$$

Then, for a system of two electrons, one might suspect that the wave function

$$\varphi_{1s}(1)\varphi_{1s}(2) \equiv \varphi_{1s}(\mathbf{r}_1)\varphi_{1s}(\mathbf{r}_2) \tag{13.33}$$

is admissible as it has the lowest possible energy state for the system. This satisfies the "indistinguishability" criterion, because it is unchanged when $\mathbf{r}_1$ is interchanged with $\mathbf{r}_2$, and it leads to an energy twice that of a single electron. But it is not antisymmetric. If we consider the wave function φ_{2s} of the $2s$ orbitals, the possible wave functions are

$$\varphi_{1s}(1)\varphi_{2s}(2) \quad \text{and} \quad \varphi_{1s}(2)\varphi_{2s}(1).$$

But these functions do not satisfy the indistinguishability criterion: We do not get the same function by interchanging 1 and 2. However, the following linear combinations do satisfy this criterion:

$$\Big(\varphi_{1s}(1)\varphi_{2s}(2) \pm \varphi_{1s}(2)\varphi_{2s}(1)\Big)/\sqrt{2} = \varphi_{\pm}^{1,2}. \tag{13.34}$$

The three spatial functions in (13.33) and (13.34) do satisfy the criterion that they are unchanged when electrons 1 and 2 are interchanged, but they are not all antisymmetric; indeed, only $\varphi_{-}^{1,2}$ is antisymmetric: $\varphi_{-}^{1,2}(1,2) = -\varphi_{-}^{1,2}(2,1)$. But we have yet to take into account spin.

Recall that α and β are the electron spin functions with $\alpha(\frac{1}{2}) = 1$, $\alpha(-\frac{1}{2}) = 0$, $\beta(\frac{1}{2}) = 0$, $\beta(-\frac{1}{2}) = 1$, depending on the quantum number of the electron. For the two-electron system, there are four spin states; we will denote them by $\alpha_1 = \alpha(1)$, $\alpha_2 = \alpha(2)$, $\beta_1 = \beta(1)$, and

$\beta_2 = \beta(2)$, with α denoting the spin state corresponding to spin-quantum number $m = 1/2$ and β corresponding to $m = -1/2$ (it is understood that the notation $\alpha(1)$ refers to the α spin function of electron 1 and can be 1 or 0, depending on whether the spin at 1 is $+\frac{1}{2}$ or $-\frac{1}{2}$, etc.). The combined states for two electrons may be

$$\alpha(1)\alpha(2), \quad \beta(1)\beta(2), \quad \left(\alpha(1)\beta(2) \pm \alpha(2)\beta(1)\right)/\sqrt{2} \quad (13.35)$$

(the $1/\sqrt{2}$ for normalization). All of these are symmetric with respect to an exchange of electrons except $(\alpha(1)\beta(2) - \alpha(2)\beta(1))/\sqrt{2}$, which is antisymmetric.

But Pauli's Principle must be applied to the combined orbitals and spins: the *spin orbitals*. Consider as possibilities of the product of the spins (13.35) and the orbitals in (13.34). Only the states

$$\varphi_+^{1,2}\sigma_-, \quad \varphi_-^{1,2}\sigma_+, \quad \varphi_-^{1,2}\alpha(1)\alpha(2), \quad \text{and} \quad \varphi_-^{1,2}\beta(1)\beta(2),$$

where $\sigma_\pm = (\alpha(1)\beta(2) \pm \alpha(2)\beta(1))/\sqrt{2}$, are antisymmetric and therefore admissible as spin orbitals for electrons.

In general, wave functions that satisfy Pauli's Exclusion Principle can be written in the form of a determinant called the *Slater determinant*, which exploits the property of determinants that the sign of a determinant is changed when any pair of rows are interchanged. Thus, for two electrons in the ground ($1s$) state of helium, the wave function is

$$\psi(1,2) = \frac{1}{\sqrt{2}} \begin{vmatrix} \varphi_{1s}(\mathbf{r}_1)\alpha(1) & \varphi_{1s}(\mathbf{r}_2)\alpha(2) \\ \varphi_{1s}(\mathbf{r}_1)\beta(1) & \varphi_{1s}(\mathbf{r}_2)\beta(2) \end{vmatrix}$$

$$= \frac{1}{\sqrt{2}}\varphi_{1s}(\mathbf{r}_1)\varphi_{1s}(\mathbf{r}_2)\left[\alpha(1)\beta(2) - \alpha(2)\beta(1)\right]$$

$$= -\psi(2,1).$$

The above determinant is the Slater determinant for this choice of wave function. The functions

$$\chi_{1s}^\alpha(\mathbf{r}_1) = \varphi_{1s}(\mathbf{r}_1)\alpha_1(1) \quad \text{and} \quad \chi_{1s}^\beta(\mathbf{r}_1) = \varphi_{1s}(\mathbf{r}_1)\beta(1)$$

are spin orbitals, and $\psi(1,2)$ can be expressed as

$$\psi(1,2) = \frac{1}{\sqrt{2}} \begin{vmatrix} \chi_{1s}^{\alpha}(\mathbf{r}_1) & \chi_{1s}^{\beta}(\mathbf{r}_1) \\ \chi_{1s}^{\alpha}(\mathbf{r}_2) & \chi_{1s}^{\beta}(\mathbf{r}_2) \end{vmatrix}.$$

The Slater determinant can be calculated for approximations of wave functions for atoms with any number N of electrons. Denoting the spin orbitals

$$\varphi_{1s}(\mathbf{r}_1)\alpha(1) = 1s(1) \quad \text{and} \quad \varphi_{1s}(\mathbf{r}_1)\beta(1) = \overline{1s(1)},$$

the Slater determinant for beryllium is (with four electrons, two in the $1s$ orbital and two in the $2s$ orbital)

$$\psi(\mathbf{r}_1, \mathbf{r}_2, \mathbf{r}_3, \mathbf{r}_4) \approx \frac{1}{\sqrt{4!}} \begin{vmatrix} 1s(1) & \overline{1s(1)} & 2s(1) & \overline{2s(1)} \\ 1s(2) & \overline{1s(2)} & 2s(2) & \overline{2s(2)} \\ 1s(3) & \overline{1s(3)} & 2s(3) & \overline{2s(3)} \\ 1s(4) & \overline{1s(4)} & 2s(4) & \overline{2s(4)} \end{vmatrix}.$$

If any two rows are the same, the determinant has a value 0, so such a state cannot exist by Pauli's Principle. Any two electrons tend to avoid each other if they are described by antisymmetric wave functions. In effect, parallel electron spins are avoided. Once again we infer that a *maximum of two electrons can occur in an atomic orbital, each with different spins*.

The electronic structure of an atom consists of a sequence of *shells* of electron density; each shell consists of the orbitals of a given quantum number n. It is customary to refer to the shell $n = 1$ as the K shell, $n = 2$ the L shell, $n = 3$ the M shell, etc. These properties are fundamental in organizing the various atomic structures of the elements in the periodic table and will be discussed in more detail in the next chapter.

13.6 Variational Principle

The various wave functions and spin orbitals described earlier must be understood to be only approximations of wave functions for multielectron atoms. They can be used, however, to build good approximations of energy states of atoms using the Rayleigh–Ritz method.

We begin by introducing the space $\mathcal{V}$ of admissible functions,

$$\mathcal{V} = \left\{ \psi \in H^1(\mathbb{R}^{3N}) : \psi = \psi(\mathbf{r}_1, \mathbf{r}_2, \ldots, \mathbf{r}_N); \right.$$

$$\left. \psi \text{ is an antisymmetric function of } (\mathbf{r}_1, \mathbf{r}_2, \ldots, \mathbf{r}_N) \right\}. \quad (13.36)$$

The variational theorem is the following:

Theorem 13.C *Let H be the Hamiltonian of a quantum system and let E_g denote its ground state energy. Then*

$$E_g \leq \langle \psi, H\psi \rangle \qquad \forall \psi \in \mathcal{V} \text{ with } \langle \psi, \psi \rangle = 1. \quad (13.37)$$

In other words,
$$E_g \leq \langle H \rangle. \quad (13.38)$$

Proof Let $\psi = \sum_k c_k \varphi_k$, $\langle \varphi_k, \varphi_j \rangle = \delta_{ij}$. Recall that $H\varphi_k = E_k \varphi_k$. Then

$$\langle H \rangle = \langle \psi, H\psi \rangle$$
$$= \left\langle \sum_k c_k \varphi_k, H \sum_j c_j \varphi_j \right\rangle$$
$$= \sum_k \sum_j c_k^* c_j E_j \langle \varphi_k, \varphi_j \rangle$$
$$= \sum_j |c_j|^2 E_j$$
$$\geq E_g \sum_j |c_j|^2 = E_g. \qquad \square$$

This variational principle provides a basis for Rayleigh–Ritz approximations of the ground state energies for more complex atoms. Let $\mathcal{V}_M$ be a finite-dimensional subspace of $\mathcal{V}$ spanned by a set of spin orbitals that satisfy Pauli's principle for fermions. Then the Rayleigh–Ritz approximation E_g^M of E_g satisfies

$$
E_g^M = \inf_{\psi \in V_M} \frac{\langle \psi, H\psi \rangle}{\langle \psi, \psi \rangle} \geq \inf_{\psi \in V} \frac{\langle \psi, H\psi \rangle}{\langle \psi, \psi \rangle} = E_g. \tag{13.39}
$$

We mention a classical example given in Messiah [37, p. 771]: The helium atom for which the Hamiltonian, we recall, is

$$
H = -\frac{\hbar^2}{2m}(\Delta_1 + \Delta_2) - \frac{e^2}{4\pi\varepsilon_0}\left(\frac{2}{r_1} + \frac{2}{r_2} - \frac{1}{|\mathbf{r}_1 - \mathbf{r}_2|}\right).
$$

In the Rayleigh–Ritz method, we introduce a trial function,

$$
\psi_a(\mathbf{r}_1, \mathbf{r}_2) = \frac{a^{-3}}{\pi}e^{-(r_1+r_2)/a}, \tag{13.40}
$$

where a is a variational parameter. The function ψ_a is suggested by perturbation theory. The corresponding energy is

$$
E(a) = \langle \psi_a, H\psi_a \rangle / \langle \psi_a, \psi_a \rangle. \tag{13.41}
$$

We chose a so that $E(a)$ is a minimum and obtain (see Messiah [37, p. 772])

$$
E_a \approx -2\left(Z - \frac{5}{16}\right)^2 E_H, \tag{13.42}
$$

where $Z = 1/2\pi\varepsilon_0$ and E_H the ground energy of the hydrogen atom. Numerically,

$$
E_a = -76.6 \text{ eV},
$$

which compares remarkably well to the experimentally determined value of -78.6 eV $= E_{\exp}$ given in Messiah [37]. Note that $E_a > E_{\exp}$.

CHAPTER 14

ATOMIC AND MOLECULAR STRUCTURE

14.1 Introduction

In this chapter, we take a brief diversion into an elementary study of electronic structures of atoms and molecules made possible by the developments in Chapters 12 and 13. Our goal is to review the classification of chemical elements based on their electronic structure determined by quantum theory and to review how these structural properties are categorized in the periodic table. In addition, we examine the basic elements of atomic bonds and how molecules are produced by binding atoms together.

14.2 Electronic Structure of Atomic Elements

The orbitals $\psi_{n\ell m}$ of the hydrogen atom are used as building blocks for categorizing the electronic structures of all the atomic elements of matter. These are the eigenfunctions of the hydrogen Hamiltonian, and their corresponding eigenvalues are the energy states which correspond to electron shells of the atom. Some of the fundamental rules of electronic orbitals and other properties that follow directly from our developments

An Introduction to Mathematical Modeling: A Course in Mechanics, First Edition. By J. Tinsley Oden
© 2011 John Wiley & Sons, Inc. Published 2011 by John Wiley & Sons, Inc.

up to now are listed as follows:

1. We number the energy shells consecutively in correspondence with the principal quantum number n ($n = 1, 2, \ldots$).

2. The spherical orbitals (corresponding to $\ell = 0$) are labeled ns, with $1s$ in the lowest energy shell, $2s$ the next, etc. These properties can be understood by reviewing properties of the hydrogen orbitals $\psi_{n\ell m}$. For $n = 1, \ell = 0, m = 0$, we have the $1s$ orbital:

$$\psi_{100} = \frac{1}{\sqrt{\pi a_0^3}} \, e^{-r/a_0}, \qquad a_0 = \frac{4\pi\epsilon_0 \hbar^2}{me^2},$$

with ground state energy

$$E_1 = -\frac{m}{2\hbar^2} \left(\frac{e^2}{4\pi\epsilon_0} \right)^2 = -13.6 \text{ eV}.$$

For $n = 2$, we have exactly four orbitals:

$$\psi_{200}, \qquad \psi_{21(-1)}, \qquad \psi_{210}, \qquad \psi_{211}.$$

Linear combinations of the last three yield the orbitals $2p_x$, $2p_y$, and $2p_z$. Thus the degeneracy of the second energy shell is 4, and

$$E_2 = -\frac{13.6}{2^2} \text{ eV} = -3.4 \text{ eV} > E_1.$$

Continuing in this way, we determine that the degeneracy of the energy level E_n is n^2. Recall that $m = -\ell, -\ell + 1, \ldots, \ell$; so for given ℓ there are $2\ell + 1$ values of m. But $\ell = 0, \ldots, n - 1$. So there are n values taken on by ℓ. So in the nth shell, there are $\sum_{\ell=0}^{n-1} (1 + 2\ell) = n^2$. Thus, there are n^2 orbitals "in" the nth energy shell, all with the same energy E_n.

3. The so-called p orbitals occur at energy level 2 and higher (only s orbitals occur at energy level 1 because $\ell = 0, \ldots, n - 1$) and they always occur in three mutually-orthogonal components, p_x, p_y, p_z. The p orbitals at the second energy level are labeled $2p_x$,

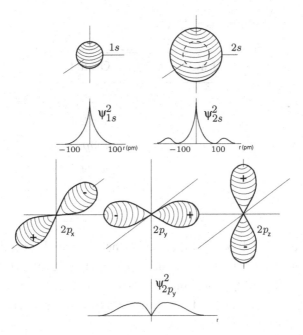

Figure 14.1: The $1s$, $2s$, and $2p$ orbitals and their corresponding probability density components, ψ^2_{1s}, ψ^2_{2s}, and $\psi^2_{2p_y}$ (cf. Gil [18, pp. 55–61]); pm denotes picometers: 10^{-12} m.

$2p_y$, $2p_z$; at the third level, $3p_x$, $3p_y$, $3p_z$, etc. *All energy levels except the first have p orbitals.*

The p orbitals occur as balloon-like lobes (see Fig. 14.1) with opposite signs of the orbital on opposite sides of the lobes. While the orbitals $1s$, $2s$, $2p$, etc., are referred to as spherical or as balloon-like, we observe that the corresponding probability density functions, ψ^2_{1s}, ψ^2_{2s}, $\psi^2_{2p_x}$, etc., are always positive and decay rapidly with the radial distance from the nucleus. Thus the convention is that the orbitals represent regions in which there is a high probability (90% or 95%) of the electron to reside.

4. At the third level, five d orbitals occur (as well as the $3s$ and $3p$ orbitals), and at level four there are an additional seven so-called f orbitals (as well as $4s$, $4p$, and $4d$ orbitals), so we have the

structure tabulated as follows:

Energy level	Orbital
1	$1s$
2	$2s, 2p$
3	$3s, 3p, 3d$
4	$4s, 4p, 4d, 4f$
$\vdots$	$\vdots$

Thus, at level 1 (shell 1), there is one spherical orbital; at level 2, there are four orbitals (1 s and 3 p lobes); at level 3, there are nine orbitals (1 s, 3 p lobes and 5 d orbitals); and at level 4, there are 16 orbitals; etc. All of these numbers of orbitals correspond to the fact that $m = -\ell, -\ell + 1, \ldots, \ell$.

5. *Each orbital can contain only (up to) two electrons.* This is because the spin momentum must be balanced: A spin up ↑ is balanced with a spin down ↓; i.e., each orbital is allowed a maximum of two electrons of opposite spin.

6. Electrons will fill low-energy orbitals (those closer to the nucleus) before the higher-energy orbitals. The s orbitals will always have a slightly lower energy than the p orbitals, for example, meaning that the s orbitals will fill before the p orbitals.

7. A universal *electronic configuration notation* is often used with the convention

$$ns^\alpha \, n'p^\beta \, n''d^\gamma \cdots \quad (n, n', n'', \ldots = \text{energy level}, \quad \alpha, \beta, \gamma \geq 1),$$

so that each element is assigned a label defining its electronic configuration. For example, carbon has six electrons and has the particular configuration

$$1s^2 2s^2 2p_x^1 2p_y^1,$$

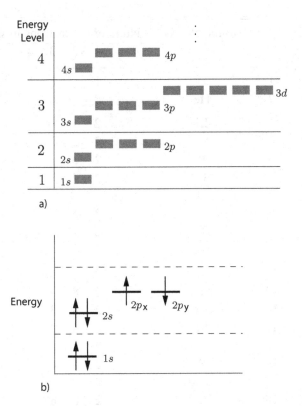

Figure 14.2: (a) Typical energy levels of various orbitals and (b) the electron configuration of carbon C(2,4) $\sim 1s^2 2s^2 2p_x^1 2p_y^1$.

meaning that there are two electrons in the $1s$ orbital, two in the $2s$ orbital, and one in the $2p_x$ and $2p_y$ orbitals; see Fig. 14.2.

14.3 The Periodic Table

Using the rules and properties we have just established, we can begin to categorize the atoms occurring in nature by their electronic structures. For example, the beginning of a table includes the following data:

Period	Z	Atomic element	Number of electrons in shell n			
			$n =$ 1	2	3	4
1	1	H	1			
	2	He	2			
2	3	Li	2	1		
	4	Be	2	2		
	5	B	2	3		
	6	C	2	4		
	7	N	2	5		
	8	O	2	6		
	9	F	2	7		
	10	Ne	2	8		
3	11	Na	2	8	1	
	12	Mg	2	8	2	
	13	Al	2	8	3	
	14	Si	2	8	4	
	15	P	2	8	5	
	16	S	2	8	6	
	17	Cl	2	8	7	
	18	Ar	2	8	8	
4	19	K	2	8	8	1
	20	Ca	2	8	8	2
$\vdots$	$\vdots$	$\vdots$	$\vdots$			

Here, Z is the number of protons in the nucleus of the atom.

These data, and much more, are convenientaly displayed in the *Periodic Table*, given in Fig. 14.3. This classical table, early forms of which go back to the eighteenth century, recognizes that all of the matter of the universe is composed of only a little over 100 basic elements (118 in our table) which differ in their number of electrons, their electronic structure, and mass. Atoms named above and displayed in the periodic table are abbreviated with one or two letters such as H for hydrogen, He for helium, etc. The number of electrons in each element is its atomic number, and the number of energy shells of an atom is called its period (thus, the "periodic table"). The table also includes the atomic mass of each element, which is a multiple of the number of protons and the average number of neutrons in its nucleus (for a given element, the number of neutrons can vary, and these are called isotopes of the element). The table is organized so that its rows correspond to the periods of the elements. Thus, elements in row one have one energy shell, those in row two have two shells, etc. Thus, there are seven rows in the table, generally labeled K, L, M, N, O, P, Q. The columns of the table are called *groups*. In general, elements in a group have the same number of electrons in their outer orbit, but there are exceptions (helium, for example, has two electrons in its outer shell, but is listed in group 8). Sodium, for example, is in period 3 (three shells), group 1 (one electron in its outer shell), and has atomic number 11 (11 electrons, 2 in shell K, 8 in shell L, and only 1 in shell M). Chlorine is in period 3, group 7a: 3 shells, 17 electrons (2 in shell K, 8 in shell L, and 7 in shell M).

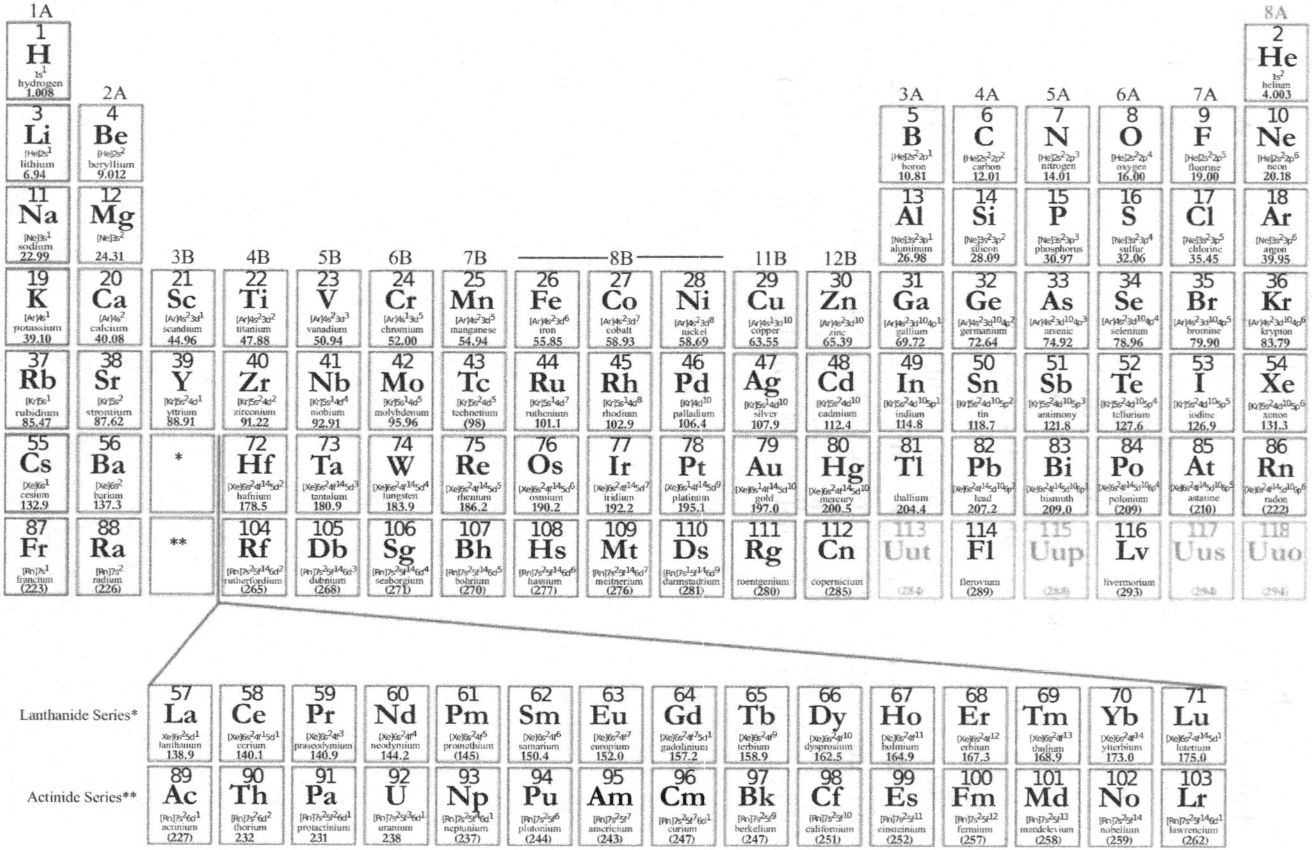

Figure 14.3: The Periodic Table. [Reproduced with permission of Los Alamos National Laboratory.]

Lanthanide Series*

Actinide Series**

14.4 Atomic Bonds and Molecules

When two or more atoms are held strongly together, we call the aggregate a *molecule*; and the forces that hold the aggregate together are called *chemical* or *atomic bonds*. Each atom has a characteristic ability to combine with others called its *valence*, which measures the number of bonds it can form. Generally speaking, the valence of an atom is the number of electrons that are needed to fill its outer shell. The hydrogen atom, for example, has one "free" electron and is assigned a valence of 1. It is said to be "univalent." Oxygen, which can form bonds with two hydrogen atoms to produce H_2O, has a valence of 2. Thus, the number of electrons in the outside shell of an atom determines the number of available electrons and, therefore, the valence of the atom.

The basic principle behind atomic bonding is that (a) atomic structures of atoms are most stable when they have no partially filled electron shells and (b) atoms will share one or more electrons to fill shells and arrive at a more stable, lower combined energy state. There are several classifications of bonds, and we will describe the most common structures as follows.

Ionic Bonds   Ionic or electrovalent bonding is an intermolecular chemical bonding in which one or more electrons leave one atom to attach to another, forming two ions of opposite charge, generating an electrical force that holds the atoms together. The donor atom will take on a positive charge and become a positive *ion*, the acceptor atom will assume a negative charge (becoming a negative ion), and the combined atoms will have a neutral charge.

The most common example is table salt: NaCl. Sodium has an atomic number of 11, with two electrons in $1s$, eight in shell 2 $(2s,2p_x,2p_y,2p_z)$, and one in the outer shell $3s$ $\big($written Na(2,8,1)$\big)$. The sodium ion is produced when the electron in the outer shell is removed, producing $Na^+ = Na(2,8,-)$. Chlorine has 17 electrons: Cl(2,8,7), and $Cl^- = Cl(2,8,8)$. Thus, an ionic bond is created when sodium and chlorine ions are bonded together to create sodium chloride NaCl. This is illustrated schematically in Fig. 14.4.

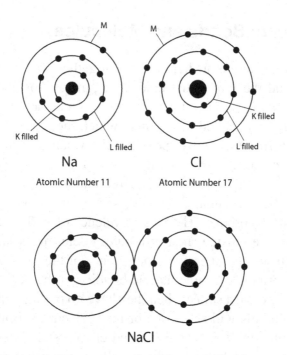

Figure 14.4: Bohr representation of ionic bonding of sodium and chlorine.

Other examples of ionic bonding are magnesium oxide:

$$Mg(2, 8, 2) + O(2, 6) \longrightarrow MgO,$$

taking 2 electrons from the outer shell of magnesium Mg and adding them to the outer shell of oxygen O to complete the configuration (2,8) so that the shells are complete. Calcium chloride involves one calcium atom, Ca(2,8,8,2), and two chlorine atoms, Cl(2,8,7), to produce $CaCl_2$.

Symbolically, the last two bonds are depicted as

$$\begin{matrix} Mg(2, 8, 2) \\ O(2, 6) \end{matrix} \implies \begin{matrix} Mg^{2+}(2, 8, -) \\ O^{2-}(2, 8, -) \end{matrix} \implies MgO,$$

$$\begin{matrix} Cl(2, 8, 7) \\ Ca(2, 8, 8, 2) \\ Cl(2, 8, 7) \end{matrix} \implies \begin{matrix} Cl^{-}(2, 8, 8, -) \\ Ca^{2+}(2, 8, 8, -) \\ Cl^{-}(2, 8, 8, -) \end{matrix} \implies CaCl_2.$$

It is common practice to illustrate these bonding transfers by means of elementary Bohr-type models, as shown in Fig. 14.4, or symbolically, as a combining of ionized atoms:

$$Na^+Cl^- = NaCl,$$
$$Cl^-Mg^{2+}Cl^- = MgCl_2,$$
$$Cl^-Ca^{2+}Cl^- = CaCl_2.$$

Covalent Bonds   Covalent bonding is another form of intermolecular bonding in which one or more pairs of electrons are shared by atoms to complete their outer shell. Covalent bonds may be assigned a direction or orientation because the bonded atoms tend to remain in fixed position relative to one another.

When two atoms with incomplete shells are in the proximity of one another, the orbitals of the outside shell may overlap forming a bonding molecule orbital and an anti-bonding orbital (where repellant forces may prevail). A stable covalent bond is formed when the bonding orbital is filled with two electrons and the anti-bonding orbital is empty. This full-bonding orbital, of course, is more stable than the original orbital. When there is a "head-on" overlap of two atomic orbitals, resulting in an intersection with a nearly circular cross section, the bond is sometimes called a sigma bond and is the strongest covalent bond.

One of the most common examples of a molecule involving covalent bonding is methane: one carbon atom $C(2,4)$ bonded with four hydrogen atoms $H(1)$; depicted by the standard symbolism, with a dash suggesting a bond and a shared electron:

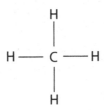

The more cumbersome but explicit Bohr model is given in Fig. 14.5.

The diagrams depicting molecular structure such as that above are called line diagrams, with a line depicting each bond. The valence of

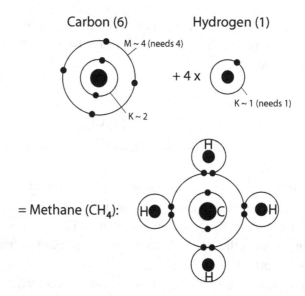

Figure 14.5: Covalent bonding of a carbon atom and four hydrogen atoms to create methane.

an atom can be represented by one to eight dots around the letter for the atom and dashes indicate covalent bonds. In classical notation, we give as examples, ethene:

$$\begin{array}{c} H \\ \diagdown \\ \end{array} C = C \begin{array}{c} H \\ \diagup \\ \end{array} \quad \sim C_2H_4$$

or ethyne:

$$H - C \equiv C - H \quad \sim C_2H_2$$

or benzene:

or we write the bonding as a reaction. An example is ammonia H_3N (with valence 2) combining with boron trichloride BCl_3:

$$H_3N: + BCl_3 \longrightarrow H_3N–BCl_3,$$

the dash indicating a covalent bond.

Metallic Bonds   In metals, which generally contain only one to three electrons in their outer shell, the outer electrons leave the atoms and become a common electron cloud, only weakly bonded to the nuclei of the atoms making up a solid. These electrons have mobility and enable the conduction of heat or electricity.

Hydrogen Bonds   In covalently bonded molecules which contain hydrogen, such as H_2O, the electrons shared by covalent bonds tend to congregate more away from the hydrogen atoms, leaving a small positive charge around the hydrogen atom and a small negative charge around its donor (e.g. oxygen or nitrogen). The negatively charged portion of one molecule will be attracted to the positively charged end, creating a hydrogen bond. A typical schematic of the situation is shown in Fig. 14.6.

The hydrogen bond is thus a bond between molecules as opposed to atoms. This is an example of a more general situation called a *polar bond*, in which two atoms covalently bonded share electrons unevenly, resulting in one part of the molecule attracting more electrons on average and thereby acquiring a slightly negative charge, the other part acquiring a slightly positive charge. The atom attracting the greatest number of electrons is said to have a higher *electronegativity* than the other atom.

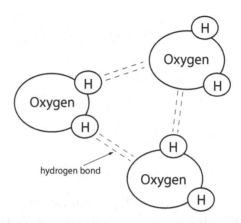

Figure 14.6: Example of hydrogen bonds involving three water molecules.

Van der Waals Bonds The van der Waals bond, also called van der Waals interaction, is a weak, secondary intermolecular bonding that arises from atomic or molecular charge distributions or dipoles, occurring due to the fluctuating polarizations of nearby electrons. These occur when there is separation between positive or negative charges in an atom or molecule.

The covalent, ionic, and metallic bonds are called *primary bonds* while van der Waals and hydrogen bonds are *secondary bonds*.

Hybridization In many compounds, orbitals become mixed in a process called *hybridization*. Typically, the $2s$ and $2p$ orbitals mix to form three types of orbitals:

- sp^3 hybridization, in which $2s$ is mixed with all three $2p$ orbitals;

- sp^2 hybridization, in which $2s$ is mixed with two $2p$ orbitals;

- sp hybridization, in which $2s$ is mixed with one $2p$ orbital.

By "mixing", we mean that a linear combination of orbitals $2s + \lambda 2p_x$, etc., is produced which distorts the orbital shapes to accommodate stable electronic configurations of certain compounds.

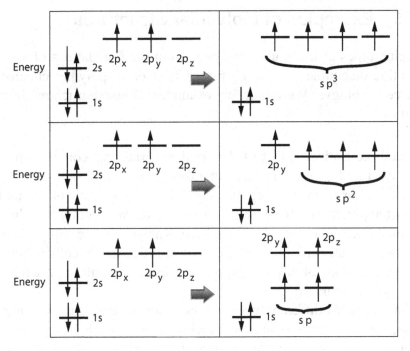

Figure 14.7: Hybrid orbitals of the carbon atom.

The energy levels of the sp^3 hybridization are depicted in Fig. 14.7, with the elongated orbital, which has a major and minor lobe (see Fig. 14.8). The methane molecule described earlier is actually created through the three-dimensional structure shown, with the four hydrogen atoms forming a strong σ-bond with four lobes of the sp^3 orbital along axes of a tetrahedral structure. The sp^2 hybridization results in a three-lobed orbital, as shown in Fig. 14.8, and provides the bonding structure for the creation of alkines or the ethane molecule introduced earlier. In sp hybridization, one has two half-filled $2p$ orbitals and two half-filled sp orbitals. This structure accommodates bonding in molecules such as ethyne.

14.5 Examples of Molecular Structures

Specific types of molecular structures form the foundations of several scientific disciplines, including materials science, polymer chemistry, and cell biology. We cite a few examples of such structures in this section.

Crystalline Solids   Most of the solid materials we come in contact with on a daily basis—metals, limestone, sand, clay, salt, etc.—have crystalline structures: a regular, repeating arrangement of atoms (at room temperature). In crystals, atoms are packed tightly together in minimal-energy configurations and are arranged in repeating clusters of atoms metallically or covalently bonded together in cells to form a lattice. The study of physical properties of such crystalline structures is called crystallography.

In matter, 14 different types of crystal structures occur, and, among these, the simplest are cubic units with atoms at each corner. In this category, additional atoms may be located within the cubic unit to give it stability. Most metals have cubic lattice structures and have one of the following forms: cubic, body-centered-cubic (bcc) with atoms located internal to the cube, or face-centered-cubic (fcc) with atoms in the side faces of the cube. Other crystalline solids have more complex lattice structures such as the hexagonal close packed (hcp), which is hexagonal in shape with atoms along diagonals. Schematics of these lattice molecules are given in Fig. 14.9.

Amorphous Solids   A solid material composed of molecules with its atoms held apart, generally in a specific pattern but with no overall periodicity, is called an amorphous solid. Examples are glass and certain plastics. Figure 14.10 shows SiO_2 (quartz), which is an amorphous solid.

Cell-Biology Molecules — DNA   The most important molecule in human existence is DNA: *deoxyribose nucleic acid*. This is a pair of long intertwined molecules (the lengths are 200,000 times the widths), each composed of a chain of deoxyribose sugar units linked to an oxygen–phosphorus molecule called the *phosphodiester linkage*. The bonded sugar–phosphodiester pair binds with a base acid, and the resulting molecule is called a *nucleotide*; the atomic structure of phosphodiesters, which is shown in Fig. 14.11, forms the *backbone* of the molecule. Each sugar unit binds to one of four possible base acid molecules: T, thymine; A, adenine; C, cytonine; and G, guanine. These structures are given in Fig. 14.12. Base pairing is stronger between A and T and between C and G. The two strands of molecules bind together as the attracting base pairs bend to form the famous helix, shown in Fig. 14.13.

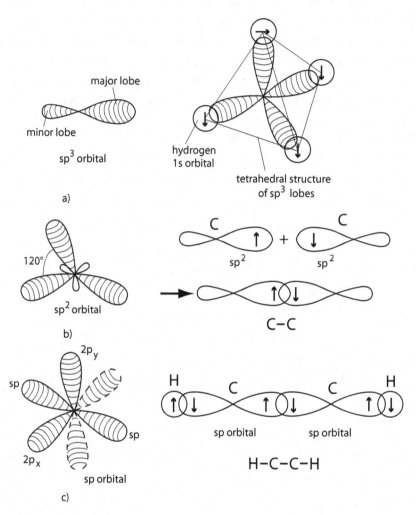

Figure 14.8: Bond structure of carbon-hydrogen molecules with hybrid orbitals.

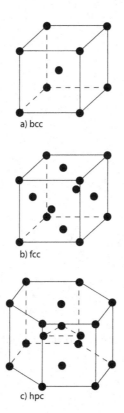

a) bcc

b) fcc

c) hpc

Figure 14.9: Crystalline cubic structures: (a) body-centered cubic, (b) face-centered cubic, and (c) hexagonal close packed crystals.

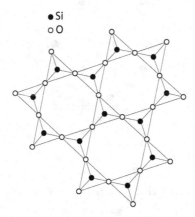

Figure 14.10: The SiO_2 (quartz) molecular structure.

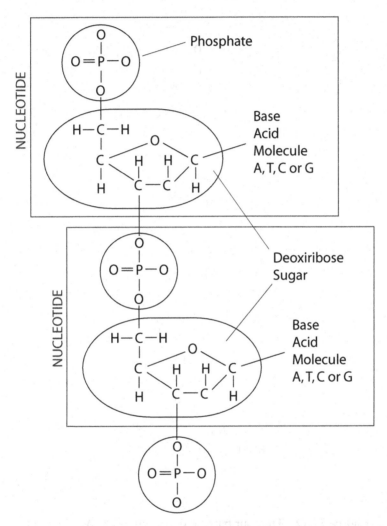

Figure 14.11: DNA nucleotide molecules consisting of deoxyribose sugar and phosphate units binding to one of the four base acid molecules.

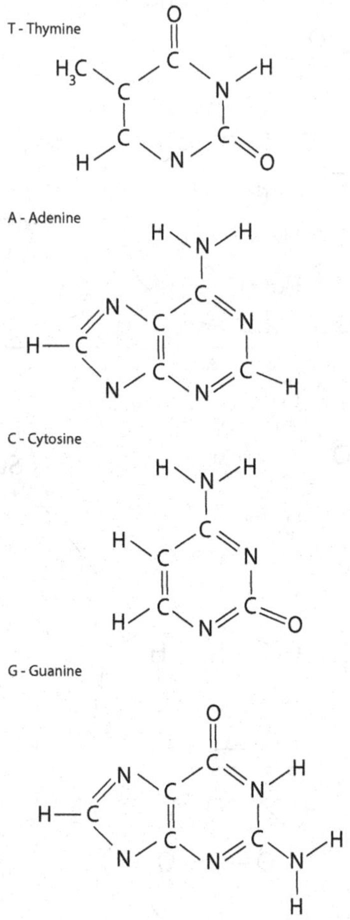

Figure 14.12: The four base acid molecules: T, A, C, and G.

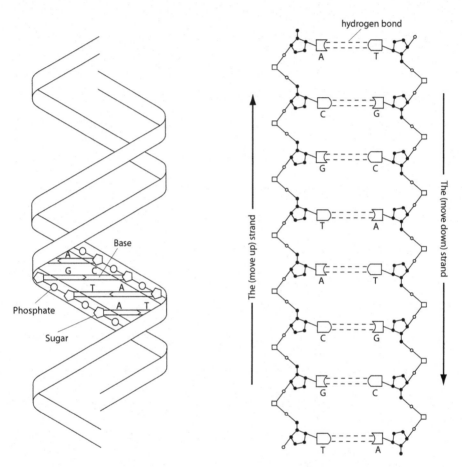

Figure 14.13: The DNA double helix, showing binding of base acids.

CHAPTER 15

AB INITIO METHODS: APPROXIMATE METHODS AND DENSITY FUNCTIONAL THEORY

15.1 Introduction

In this chapter we will embark on an introductory study of several powerful methods for studying quantum systems with many atoms and electrons. These methods are often referred to as *ab initio* methods, meaning "from first principles" or "from the beginning." But all involve assumptions, approximations, or empirical information. For instance, we base all calculations of electronic properties covered here on the Born–Oppenheimer approximation, discussed in the next section. Indeed, since it is quite impossible to solve Schrödinger's equation exactly for the wave function of such systems, we seek good approximations of ground state energies using the Rayleigh–Ritz approach. We discuss several fundamental wave function methods for such systems, such as the Hartree–Fock theory. But importantly, we also provide an introduction to Density Functional Theory, which provides a powerful alternative to

An Introduction to Mathematical Modeling: A Course in Mechanics, First Edition. By J. Tinsley Oden © 2011 John Wiley & Sons, Inc. Published 2011 by John Wiley & Sons, Inc.

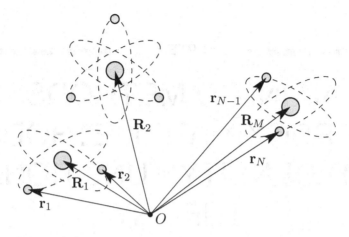

Figure 15.1: A molecular system consisting of M nuclei and N electrons.

wave function methods based on the notion of electron density.

15.2 The Born–Oppenheimer Approximation

We now wish to consider a general molecular system made up of M atoms and a total number of N electrons, such as that indicated in Fig. 15.1. This is an $(N + M)$-body problem. The M nuclei are located by position vectors $\mathbf{R}_1, \mathbf{R}_2, \ldots, \mathbf{R}_M$ emanating from an origin O, and the N electrons are located by the position vectors $\mathbf{r}_1, \mathbf{r}_2, \ldots, \mathbf{r}_N$. The nuclei have charges $Z_K e$, $K = 1, 2, \ldots, M$. The nonrelativistic Hamiltonian for this general molecular system is

$$
\begin{aligned}
H = {} & -\frac{\hbar^2}{2m} \sum_{i=1}^{N} \Delta_{\mathbf{r}_i} - \frac{\hbar^2}{2} \sum_{K=1}^{M} \frac{1}{M_K} \Delta_{\mathbf{R}_K} \\
& + \sum_{K=1}^{M} \left(-\sum_{i=1}^{N} \frac{Z_K e^2}{4\pi\epsilon_0 r_{Ki}} + \sum_{L>K} \frac{Z_K Z_L e^2}{4\pi\epsilon_0 R_{KL}} \right) + \sum_{i=1}^{N} \sum_{j>i}^{N} \frac{e^2}{4\pi\epsilon_0 r_{ij}}.
\end{aligned}
$$

$$(15.1)$$

Here we employ the usual notations:

m = the electron mass,

M_K = the mass of nucleus K,

eZ_K = the charge of nucleus K,

e = the electron charge,

$$\Delta \mathbf{r}_i = \frac{\partial^2}{\partial r_{i1}^2} + \frac{\partial^2}{\partial r_{i2}^2} + \frac{\partial^2}{\partial r_{i3}^2}, \qquad \Delta \mathbf{R}_K = \frac{\partial^2}{\partial R_{K1}^2} + \frac{\partial^2}{\partial R_{K2}^2} + \frac{\partial^2}{\partial R_{K3}^2},$$

$$r_{Ki} = |\mathbf{R}_K - \mathbf{r}_i| = \left\{ \sum_{j=1}^{3} |R_{Kj} - r_{ij}|^2 \right\}^{1/2},$$

$$R_{KL} = |\mathbf{R}_K - \mathbf{R}_L|, \qquad 1 \le K, L \le M,$$

$$r_{ij} = |\mathbf{r}_i - \mathbf{r}_j|, \qquad 1 \le i, j \le N.$$

We wish to solve the time-independent Schrödinger equation,

$$H\psi = E\psi,$$

for the wave function and the energy levels E, particularly the ground state energy E_g. Denoting

$$\mathbf{r}^N = (\mathbf{r}_1, \mathbf{r}_2, \ldots, \mathbf{r}_N),$$

$$\mathbf{R}^M = (\mathbf{R}_1, \mathbf{R}_2, \ldots, \mathbf{R}_M),$$

the Hamiltonian in (15.1) can be written concisely as

$$H = T_e(\mathbf{r}^N) + T_M(\mathbf{R}^M) + V_{eM}(\mathbf{r}^N, \mathbf{R}^M) + V_{MM}(\mathbf{R}^M) + V_{ee}(\mathbf{r}^N),$$
$$(15.2)$$

where T_e and T_M are the electronic and nuclear kinetic energy terms (the first two terms on the right-hand side of (15.1)), V_{eM} is the coupled electronic–nuclear potential, V_{MM} is the nuclear Coulomb potential, and V_{ee}, the last term in (15.1), is the electronic Coulomb potential.

Specifically,

$$T_e(\mathbf{r}^N) = -\frac{\hbar^2}{2m} \sum_{i=1}^{N} \Delta_{\mathbf{r}_i},$$

$$T_M(\mathbf{R}^M) = -\frac{\hbar^2}{2} \sum_{K=1}^{M} \frac{1}{M_K} \Delta_{\mathbf{R}_K},$$

$$V_{eM}(\mathbf{r}^N, \mathbf{R}^M) = -\sum_{K=1}^{M} \sum_{i=1}^{N} \frac{Z_K e^2}{4\pi\epsilon_0 r_{Ki}}, \qquad (15.3)$$

$$V_{MM}(\mathbf{R}^M) = \sum_{K=1}^{M} \sum_{L>K}^{M} \frac{Z_K Z_L e^2}{4\pi\epsilon_0 R_{KL}},$$

$$V_{ee}(\mathbf{r}^N) = \sum_{i=1}^{N} \sum_{j>i}^{N} \frac{e^2}{4\pi\epsilon_0 r_{ij}}.$$

The wave function,

$$\psi(\mathbf{r}_1, \mathbf{r}_2, \ldots, \mathbf{r}_N, \mathbf{R}_1, \mathbf{R}_2, \ldots, \mathbf{R}_M) = \psi(\mathbf{r}^N, \mathbf{R}^M),$$

cannot be separated into products of electronic and nuclear wave functions (such as $\psi = \psi_1(\mathbf{r}^N)\psi_2(\mathbf{R}^M)$) because of the term $V_{eM}(\mathbf{r}^N, \mathbf{R}^M)$. Fortunately, a simplifying and generally valid assumption can be made at this point which allows us to separate out the electronic components of the Hamiltonian:

The Born–Oppenheimer Approximation *The nuclei are much more massive than the electrons and are essentially fixed in space. The nuclei coordinates $\mathbf{R}^M$ can be treated as parameters.*

To employ this approximation, we write

$$\psi(\mathbf{r}^N, \mathbf{R}^M) = \psi_e(\mathbf{r}^N, \mathbf{R}^M)\chi(\mathbf{R}^M), \qquad (15.4)$$

where ψ_e is the electronic wave function and χ is the nuclear wave function. Introducing this expression into the time-independent Schrödinger equation $H\psi = E\psi$, we shall assume that the nuclei follow a trajectory τ defined by varying the parameter set $\mathbf{R}^M$ such that

$$\langle \psi, \psi \rangle_r \equiv \int_{\mathbb{R}^{3N}} \psi_e^* \psi_e \, d^{3N} r = 1 \qquad \forall \mathbf{R}^M \in \tau,$$

and that also

$$\langle \chi, \chi \rangle_R \equiv \int_{\mathbb{R}^{3M}} \chi^* \chi \, d^{3M} R = 1.$$

Then, for given $\mathbf{R}^M$, we have

$$E\psi_e \chi = H_{\text{elec}} \psi_e \chi + H_{\text{nuc}} \psi_e \chi, \tag{15.5}$$

where H_{elec} and H_{nuc} are the electronic and nuclear Hamiltonians,

$$
\begin{aligned}
H_{\text{elec}} &= T_e(\mathbf{r}^N) + V_{eM}(\mathbf{r}^N, \mathbf{R}^M) + V_{ee}(\mathbf{r}^N), \\
H_{\text{nuc}} &= T_M(\mathbf{R}^M) + V_{MM}(\mathbf{R}^M),
\end{aligned}
\tag{15.6}
$$

and $\psi_e = \psi_e(\mathbf{r}^N, \mathbf{R}^M)$. The *electronic Schrödinger equation* is

$$\boxed{E_{\text{elec}}(\mathbf{R}^M)\psi_e(\mathbf{r}^N, \mathbf{R}^M) = H_{\text{elec}} \psi_e(\mathbf{r}^N, \mathbf{R}^M).} \tag{15.7}$$

Under simplifying assumptions on the separability of the electronic energy, one can argue from (15.5) that

$$E\chi = E_{\text{elec}}(\mathbf{R}^M)\chi + H_{\text{nuc}} \chi. \tag{15.8}$$

Hence, the *nuclear Schrödinger equation* can be written

$$\boxed{E\chi = \Big(T_M(\mathbf{R}^M) + V_{MM}(\mathbf{R}^M) + E_{\text{elec}}(\mathbf{R}^M)\Big)\chi.} \tag{15.9}$$

The solution to the electronic eigenvalue problem (15.7) yields an energy $E_{\text{elec}}(\mathbf{R}^M)$ that depends on the parameter $\mathbf{R}^M$, and the function

$E_{\text{elec}}(\mathbf{R}^M)$ defines what is sometimes called a *Born–Oppenheimer sur-face*. There will be one such surface for each energy level E_{elec}^n. In view of (15.9), the internuclear potential for given $\mathbf{R}^M$ is

$$V_M(\mathbf{R}^M) = V_{MM}(\mathbf{R}^M) + E_{\text{elec}}(\mathbf{R}^M). \qquad (15.10)$$

For given surfaces $E_{\text{elec}}(\mathbf{R}^M)$, the nuclear dynamics on the surface may be described by a time-dependent Schrödinger equation for a wave function $\Psi(\mathbf{R}^M, t)$:

$$i\hbar \frac{\partial}{\partial t} \Psi(\mathbf{R}^M, t) = \Big(T_M(\mathbf{R}^M) + V_M(\mathbf{R}^M) \Big) \Psi(\mathbf{R}^M, t).$$

As the nuclear system evolves in time, the electronic system, by com-parison, responds essentially spontaneously, and the electronic system can be solved with each fixed value of the parameters $\mathbf{R}^M$.

Virtually every calculation in multi-atom, molecular quantum me-chanics is based on the Born–Oppenheimer approximation, and we shall employ it throughout all developments that follow in this chapter. We shall generally assume that the nuclei are fixed in space and that only electronic properties are of interest, so that the Hamiltonian of the system is the electronic Hamiltonian of (15.7). Thus, hereafter, by the Hamilto-nian H we shall mean $H = H_{\text{elec}} = T_e + V_{eM} + V_{ee}$, by the ground state energy we shall mean $E_g = E_{\text{elec}}$, and the nuclei coordinates are fixed parameters.

15.3   The Hartree and the Hartree–Fock Methods

The use of Rayleigh–Ritz approximations of the variational problem for the ground state energy of multi-atom systems is referred to as the Hartree method in recognition of the 1928 paper on this subject by Hartree [23]. An amendment to the approach, presumably an improvement, was con-tributed shortly thereafter by Fock [17] resulting in the Hartree–Fock method. These methods are in a class of approximate methods called *wave-function methods*.

With the Born–Oppenheimer assumption in place, we have

$$H = -\frac{\hbar^2}{2m}\sum_{i=1}^{N}\Delta_{\mathbf{r}_i} - \sum_{i=1}^{N}\sum_{K=1}^{M}\frac{Z_K e^2}{4\pi\epsilon_0 r_{Ki}} + \sum_{i=1}^{N}\sum_{j>i}^{N}\frac{e^2}{4\pi\epsilon_0 r_{ij}}$$

$$= T_e(\mathbf{r}^N) + V_{eM}(\mathbf{r}^N) + V_{ee}(\mathbf{r}^N)$$

$$= H_{\text{elec}}, \tag{15.11}$$

with $V_{eM}(\mathbf{r}^N, \mathbf{R}^M) = V_{eM}(\mathbf{r}^N)$. We next introduce some simplifying notation. Let $v = v(\mathbf{r})$ denote the potential field corresponding to fixed nuclei with given charges $Z_K e$:

$$v(\mathbf{r}) = -\sum_{K=1}^{M}\frac{Z_K e^2}{4\pi\epsilon_0|\mathbf{r} - \mathbf{R}_K|}, \tag{15.12}$$

and denote by $h = h(\mathbf{r})$ the operator

$$h(\mathbf{r}) = -\frac{\hbar^2}{2m}\Delta_{\mathbf{r}} + v(\mathbf{r}). \tag{15.13}$$

In addition, let

$$U(\mathbf{r}, \mathbf{r}') = \frac{e^2}{4\pi\epsilon_0|\mathbf{r} - \mathbf{r}'|}. \tag{15.14}$$

Then the (electronic) Hamiltonian can be written

$$H = \sum_{i=1}^{N}\left(h(\mathbf{r}_i) + \sum_{j>i}^{N}U(\mathbf{r}_i, \mathbf{r}_j)\right). \tag{15.15}$$

We see that $h(\mathbf{r}_i)$ is the one-electron noninteracting Hamiltonian for electron i and that the Hamiltonian for N noninteracting electrons is

$$H_{\text{non}} = \sum_{i=1}^{N}h(\mathbf{r}_i). \tag{15.16}$$

15.3.1 The Hartree Method

The basic idea behind the classical Hartree method is to use as basis functions products of the orbitals of the noninteracting system:

$$h(\mathbf{r}_i)\, \varphi_i(\mathbf{r}_i) = \epsilon_i \varphi_i(\mathbf{r}_i), \qquad (15.17)$$

and

$$\psi_{\mathrm{H}}(\mathbf{r}_1, \mathbf{r}_2, \ldots, \mathbf{r}_N) = \varphi_1(\mathbf{r}_1)\, \varphi_2(\mathbf{r}_2) \cdots \varphi_N(\mathbf{r}_N) = \prod_{i=1}^{N} \varphi_i(\mathbf{r}_i). \quad (15.18)$$

The approximation ψ_{H} is the Hartree wave function and the product of orbitals in (15.18) is called the *Hartree product*. The energy is then $\sum_{i=1}^{n} \epsilon_i$ and electron-to-electron interactions are ignored. This is called the simple form of the Hartree method. In the full Hartree method, electron–electron interactions are included in the Hamiltonian by adding the electron–electron potential. For a single electron, we use

$$\left(h(\mathbf{r}_i) + \sum_{j>i}^{N} U(\mathbf{r}_i, \mathbf{r}_j) \right) \hat{\varphi}_i(\mathbf{r}_i) = \hat{\epsilon}_i \hat{\varphi}_i(\mathbf{r}_i), \qquad i = 1, 2, \ldots, N.$$

$$(15.19)$$

Generally, equations (15.19) are solved one electron at a time with the interactions of the $N-1$ other electrons accounted for in an iterative process. The wave function is again $\psi_{\mathrm{H}} = \prod_{i=1}^{N} \hat{\psi}_i(\mathbf{r}_i)$ and the total energy is $\langle \psi_{\mathrm{H}}, \sum_{i=1}^{N} \left(h(\mathbf{r}_i) + \sum_{j>i}^{N} U(\mathbf{r}_i, \mathbf{r}_j) \right) \psi_{\mathrm{H}} \rangle$.

15.3.2 The Hartree–Fock Method

Not surprisingly, the Hartree method may result in an inaccurate approximation of the wave function because it does not respect the Pauli principle: When two electrons occupy the same position, the wave function must vanish, and it must be an antisymmetric function of electron positions. This defect is addressed in the Hartree–Fock method, which employs antisymmetric basis functions generated using Slater determinants. Toward this end, in addition to the spatial coordinates $\mathbf{r}_i$ of each electron, we supply a spin coordinate $\sigma = \alpha$ or $\sigma = \beta$, where α and β are the spin coordinates, and denote $(\mathbf{r}, \sigma) = \mathbf{x}$ discussed in Chapter 13 The spin orbitals are now the products of orthonormal one-particle wave functions and spins, denoted as

$$\chi_i(\mathbf{x}) = \chi_i\big((\mathbf{r}, \sigma)\big) = \varphi_i(\mathbf{r})\sigma_i(s), \qquad (15.20)$$

with it understood that s is the spin of the electron at r. We demand that the Pauli principle be satisfied. So to ensure antisymmetry of the approximate wave function, we use (15.20) to compute the global electronic wave function as a Slater determinant:

$$\psi_{\mathrm{HF}} = \frac{1}{\sqrt{N!}} \begin{vmatrix} \chi_1(\mathbf{x}_1) & \chi_2(\mathbf{x}_1) & \cdots & \chi_N(\mathbf{x}_1) \\ \chi_1(\mathbf{x}_2) & \chi_2(\mathbf{x}_2) & \cdots & \chi_N(\mathbf{x}_2) \\ \vdots & \vdots & \ddots & \vdots \\ \chi_1(\mathbf{x}_N) & \chi_2(\mathbf{x}_N) & \cdots & \chi_N(\mathbf{x}_N) \end{vmatrix}. \qquad (15.21)$$

We recall that use of the Slater determinant for a trial wave function leads to an antisymmetric property of fermions and satisfies the indistinguishability requirement by vanishing if two electrons are identical ($\mathbf{x}_i = \mathbf{x}_j$). The resulting wave function ψ_{HF} can lead to an upper bound of the true electronic ground state energy (by virtue of (13.39)). It remains to be shown how the orbitals φ_i are calculated.

The energy of the Hartree–Fock approximation is

$$E_{\mathrm{HF}} = \langle \psi_{\mathrm{HF}}, H\psi_{\mathrm{HF}} \rangle$$

$$= \sum_{s_1,\ldots,s_N} \sum_{i=1}^{N} \int_{\mathbb{R}^{3N}} \frac{1}{N!} \det \left| \chi_i^*(\mathbf{x}_j) \right| \Big(h(\mathbf{r}_i) $$

$$+ \sum_{j>i} U(\mathbf{r}_i, \mathbf{r}_j) \Big) \det \left| \psi_i(\mathbf{x}_j) \right| d^{3N}r. \qquad (15.22)$$

In the Hartree–Fock method, one minimizes E_{HF} over the space of spatial orbitals, subject to the constraint that $\int_{\mathbb{R}^{3N}} \psi_{\mathrm{HF}}^* \psi_{\mathrm{HF}} \, d^{3N}r = 1$ to produce a one-electron version of Schrödinger's equation. After considerable algebra (see, e.g., Marder [36], Leach [33], or Ratner and Schatz [39]), we obtain the eigenvalue–eigenfunction problem

$$
\epsilon_i \varphi_i(\mathbf{r}) = \left(h(\mathbf{r}_i) + \sum_{j=1}^{N} J_i(\varphi_j) \right) \varphi_i(\mathbf{r})
$$

$$
- \sum_{j=1}^{N} K_i(\varphi_j, \varphi_i) \varphi_i(\mathbf{r}), \qquad i = 1, 2, \ldots, N,
$$

(15.23)

where

$$
J_i(\varphi_j) = \int_{\mathbb{R}^{3N}} \frac{e^2 |\varphi_j(\mathbf{r}')|^2}{4\pi\epsilon_0 |\mathbf{r} - \mathbf{r}'|} \, d^{3N} r',
$$

$$
K_i(\varphi_j, \varphi_i) = \int_{\mathbb{R}^{3N}} \frac{e^2}{4\pi\epsilon_0 |\mathbf{r}_i - \mathbf{r}'|} \varphi_j^*(\mathbf{r}') \, \varphi_i(\mathbf{r}') \, d^{3N} r'.
$$

The system (15.23) is called the *Hartree–Fock equations*. Here $J_i(\varphi_j)$ represents the repulsion between charges i and j and is called the *Coulomb integral* and $K_i(\varphi_j, \varphi_i)$, called the *exchange integral*, arises from the antisymmetry of the wave function.

Upon solving (15.23) for the single electron orbitals, the total energy is then calculated using the relation

$$
E_{\mathrm{HF}} = \sum_{i=1}^{N} \left(\epsilon_i - \sum_{j=1}^{N} \left(2J_{ij} - K_{ij} \right) \right),
$$

with

$$
J_{ij} = \langle \varphi_i, J_i(\varphi_j) \varphi_i \rangle,
$$
$$
K_{ij} = \langle \varphi_i, K_i(\varphi_i, \varphi_j) \varphi_i \rangle.
$$

Thus, E_{HF} is not simply the sum of orbital energies as in the Hartree method: there is now the correction $(2J_{ij} - K_{ij})$.

We observe that the eigenvalue problem (15.23) is nonlinear: The solution for orbital φ_i involves the orbitals of the other electrons. A classical solution strategy to this problem, called the *self-consistent field* (SCF) approach, is to choose a set of trial solutions φ_i (or actually χ_i) and calculate the Coulomb and exchange terms. Then compute the solutions tot the resulting Hartree–Fock equations, giving a second set of solutions, which are used in calculating corrected Coulomb and exchange integrals *Coulomb integral*. This iterative process is continued, each iterate intended to yield a lower total energy, until the calculated energy remains essentially unchanged. The orbitals then are said to be self-consistent.

15.3.3   The Roothaan Equations

For each spin orbital χ_i, we use the expansion

$$\chi_i = \sum_{\alpha=1}^{K} c_{\alpha i} \hat{\chi}_\alpha, \tag{15.24}$$

where $\hat{\chi}_\alpha$ are basis functions for a finite dimensional approximation space. We denote by S the matrix

$$S_{\alpha\beta} = \int_{\mathbb{R}^3} \hat{\chi}_\alpha^* \hat{\chi}_\beta \, dr \, d\sigma, \qquad 1 \leq \alpha, \beta \leq K. \tag{15.25}$$

The orthogonality condition, $\int_{\mathbb{R}^3} \chi_i^* \chi_j \, dr \, d\sigma = \delta_{ij}, 1 \leq i, j \leq N$, leads to the relation

$$C^* S C = I, \tag{15.26}$$

where C is the matrix of coefficients in (15.24) and I is the $N \times N$ identity matrix: $C = [c_{i\alpha}]$.

The matrix D defined by

$$D = C C^*, \tag{15.27}$$

is called the *density matrix*. Introducing (15.24) into the Hartree–Fock energy expression and minimizing $E_{\mathrm{HF}}(D)$ subject to the constraints that $D^* = D$, $\mathrm{tr}(SD) = N$, $DSD = D$ yields the nonlinear system of equations in the coefficients $\{c_{i\alpha}\}$:

$$\boxed{F(D)C = \epsilon S C,} \tag{15.28}$$

where

$$F = \left\{ \int_{\mathbb{R}^3} \int_{\mathbb{R}^3} \chi_\alpha^* \mathcal{F} \chi_\beta \, dr \, d\sigma \right\}. \qquad (15.29)$$

$\mathcal{F}$ corresponds to the terms $2J_{ij} - K_{ij}$ and $\epsilon = \text{diag}(\epsilon_1, \epsilon_2, \ldots, \epsilon_N)$. Equations (15.28) are called the *Roothaan* (or the *Roothaan–Hall*) *equations*.

It is customary to attempt to solve (15.28) using a fixed-point iterative method: Pick an initial C_0, compute $F(C_0 C_0^*)$, solve (15.28) for a correction C_1, and continue until a suitable C is obtained. This iterative process does not always converge (see Cances [12]) and it may be necessary to use more sophisticated iterative schemes.

15.4   Density Functional Theory

According to W. Kohn [32], wave function methods such as the Hartree–Fock method can give excellent results for small molecular systems in which the total number of chemically reactive electrons is of order 10. Even if this number could be increased by an order of magnitude with the use of large-scale computing systems, it seems unlikely that significant progress can be made without a dramatic alternative formulation of the many-atom quantum problem. One such family of approaches may live in the notion of Density Functional Theory, in which the search for a wave function for the system is replaced by the search for an electron density, a single scalar-valued function of position in $\mathbb{R}^3$.

15.4.1   Electron Density

Consider a quantum system of N electrons, the position of each given by a position vector r_i, $i = 1, 2, \ldots, N$. The wave function ψ will be an antisymmetric function of the N vectors

$$\psi = \psi(\mathbf{r}_1, \mathbf{r}_2, \ldots, \mathbf{r}_N),$$

and the corresponding probability density function is

$$\psi^* \psi = |\psi(\mathbf{r}_1, \mathbf{r}_2, \ldots, \mathbf{r}_N)|^2 = \varrho(\mathbf{r}_1, \mathbf{r}_2, \ldots, \mathbf{r}_N).$$

Now, let $\mathbf{r}$ be the position of an arbitrary point in $\mathbb{R}^3$. The probability that electron 1 is at $\mathbf{r}$ is given by the marginal probability

$$\mathbb{P}_1(\mathbf{r}) = \int_{\mathbb{R}^{3N-3}} \varrho(\mathbf{r}, \mathbf{r}_2, \mathbf{r}_3, \mathbf{r}_4, \dots, \mathbf{r}_N)\, d^{3N-3}r.$$

Similarly, the marginal probability electron 2 will be at $\mathbf{r}$ is

$$\mathbb{P}_2(\mathbf{r}) = \int_{\mathbb{R}^{3N-3}} \varrho(\mathbf{r}_1, \mathbf{r}, \mathbf{r}_3, \mathbf{r}_4, \dots, \mathbf{r}_N)\, d^{3N-3}r,$$

and that of electron 3 being at position $\mathbf{r}$ is

$$\mathbb{P}_3(\mathbf{r}) = \int_{\mathbb{R}^{3N-3}} \varrho(\mathbf{r}_1, \mathbf{r}_2, \mathbf{r}, \mathbf{r}_4, \dots, \mathbf{r}_N)\, d^{3N-3}r,$$

etc.   Now ψ is an antisymmetric function of the N positions $\mathbf{r}_1, \mathbf{r}_2, \dots, \mathbf{r}_N$. So $|\psi|^2$ must be unchanged by a change in the coordinates of any two consecutive vectors:

$$\begin{aligned}
\varrho(\mathbf{r}, \mathbf{r}_2, \mathbf{r}_3, \mathbf{r}_4, \dots, \mathbf{r}_N) &= \varrho(\mathbf{r}_1, \mathbf{r}, \mathbf{r}_3, \mathbf{r}_4, \dots, \mathbf{r}_N) \\
&= \varrho(\mathbf{r}_1, \mathbf{r}_2, \mathbf{r}, \mathbf{r}_4, \dots, \mathbf{r}_N) \\
&= \cdots \\
&= \varrho(\mathbf{r}_1, \mathbf{r}_2, \mathbf{r}_3, \mathbf{r}_4, \dots, \mathbf{r}).
\end{aligned}$$

So the probability that any of the N electrons is at $\mathbf{r}$ is

$$n(\mathbf{r}) = \sum_{i=1}^{N} \mathbb{P}_i(\mathbf{r}), \tag{15.30}$$

or

$$n(\mathbf{r}) = N \int_{\mathbb{R}^{3N-3}} |\psi(\mathbf{r}, \mathbf{r}_2, \mathbf{r}_3, \mathbf{r}_4, \dots, \mathbf{r}_N)|^2\, d^{3N-3}r. \tag{15.31}$$

Here $d^{3N-3}r$ means $d^3r_2\, d^3r_3 \cdots d^3r_N$ and $d^3r_i = dr_{i1}\, dr_{i2}\, dr_{i3}$, $i = 2, 3, \dots, N$. The scalar field $n(\mathbf{r})$ is called the *electron density*. Clearly, its value at $\mathbf{r}$ is the collective probability of the N electrons being located at $\mathbf{r}$; or, more precisely, $\varrho(\mathbf{r})d^2r$ is the probable number of electrons in the infinitesimal volume d^3r. It is also clearly related to the charge density, as each electron is endowed with a negative charge of e. Unlike the wave function itself, the electron density can be actually measured using X-ray crystallography. We remark that we can add the effects of spin to the definition of n by replacing the $\mathbf{r}_i$ by the position–spin coordinate $\mathbf{x}_i = (\mathbf{r}_i, \sigma_i)$ and setting

$$n(\mathbf{r}) = N \sum_{s_i} \int_{\mathbb{R}^{3N-3}} |\psi(\mathbf{x}, \mathbf{x}_2, \ldots, \mathbf{x}_N)|^2 \, d^{3N-3}x.$$

Some immediate identities involving the electron density are given in the following lemma.

Lemma 15.1 *For n given by (15.31), we have*

(i)

$$\int_{\mathbb{R}^3} n(\mathbf{r}) \, dr = N. \qquad (15.32)$$

(ii) If v is the potential (15.12), we obtain

$$\left\langle \psi, \sum_{i=1}^{N} v(\mathbf{r}_i)\psi \right\rangle = \int_{\mathbb{R}^3} v(\mathbf{r})n(\mathbf{r}) \, dr. \qquad (15.33)$$

(iii) Let U be the interaction potential (15.14) and let $\psi = \psi_\mathrm{H}$ be the Hartree wave function of (15.18). Then,

$$\left\langle \psi, \sum_{i=1}^{N}\sum_{j>i} U(\mathbf{r}_i, \mathbf{r}_j)\psi \right\rangle = \frac{c}{2}\left(1 - \frac{1}{N}\right) \int_{\mathbb{R}^3} \int_{\mathbb{R}^3} \frac{n(\mathbf{r})n(\mathbf{r}')}{|\mathbf{r} - \mathbf{r}'|} \, dr \, dr',$$

$$(15.34)$$

where $c = (e^2/4\pi\epsilon_0)$.

Proof (i) follows trivially from (15.31). To arrive at (ii), we observe that

$$
\left\langle \psi, \sum_{i=1}^{N} v(\mathbf{r}_i)\psi \right\rangle = \int_{\mathbb{R}^{3N}} \psi^*\psi(\mathbf{r}_1, \mathbf{r}_2, \ldots, \mathbf{r}_N)v(\mathbf{r}_1)\, d^{3N}r
$$

$$
+ \int_{\mathbb{R}^{3N}} \psi^*\psi(\mathbf{r}_1, \mathbf{r}_2, \ldots, \mathbf{r}_N)v(\mathbf{r}_2)\, d^{3N}r
$$

$$
+ \cdots
$$

$$
+ \int_{\mathbb{R}^{3N}} \psi^*\psi(\mathbf{r}_1, \mathbf{r}_2, \ldots, \mathbf{r}_N)v(\mathbf{r}_N)\, d^{3N}r
$$

$$
= \int_{\mathbb{R}^3} v(\mathbf{r}_1)\left(\int_{\mathbb{R}^{3N-3}} \psi^*\psi(\mathbf{r}_1, \mathbf{r}_2, \ldots, \mathbf{r}_N) \right.
$$
$$
\left. dr_2\, dr_3 \cdots dr_N \right) dr_1
$$

$$
+ \int_{\mathbb{R}^3} v(\mathbf{r}_2)\left(\int_{\mathbb{R}^{3N-3}} \psi^*\psi(\mathbf{r}_1, \mathbf{r}_2, \ldots, \mathbf{r}_N) \right.
$$
$$
\left. dr_1\, dr_3 \cdots dr_N \right) dr_2
$$

$$
+ \cdots
$$

$$
+ \int_{\mathbb{R}^3} v(\mathbf{r}_N)\left(\int_{\mathbb{R}^{3N-3}} \psi^*\psi(\mathbf{r}_1, \mathbf{r}_2, \ldots, \mathbf{r}_N) \right.
$$
$$
\left. dr_1\, dr_2 \cdots dr_{N-1} \right) dr_N.
$$

Then

$$
\left\langle \psi, \sum_{i=1}^{N} v(\mathbf{r}_i)\psi \right\rangle = \int_{\mathbb{R}^3} v(\mathbf{r}_1)n(\mathbf{r}_1)\frac{1}{N}\, dr_1 + \int_{\mathbb{R}^3} v(\mathbf{r}_2)n(\mathbf{r}_2)\frac{1}{N}\, dr_2
$$

$$
+ \cdots + \int_{\mathbb{R}^3} v(\mathbf{r}_N)n(\mathbf{r}_N)\frac{1}{N}\, dr_N
$$

$$
= \int_{\mathbb{R}^3} v(\mathbf{r})n(\mathbf{r})\, dr.
$$

(iii) Noting that $\int_{\mathbb{R}^{3N}} |\psi|^2 \, d^{3N} r = 1$, we have

$$\left\langle \psi, \sum_{i=1}^{N} \sum_{j>i}^{N} U(\mathbf{r}_i, \mathbf{r}_j) \psi \right\rangle$$

$$= \int_{\mathbb{R}^{3N}} \left(\sum_{i=1}^{N} \sum_{j>i}^{N} \frac{c}{|\mathbf{r}_i - \mathbf{r}_j|} \psi^* \psi(\mathbf{r}^N) \int_{\mathbb{R}^{3N}} \psi^* \psi(\mathbf{u}^N) \, d^{3N} u \right) d^{3N} r.$$

The first term in the sum $(i = 1, j = 2)$ is then

$$c \int_{\mathbb{R}^3} \int_{\mathbb{R}^3} \frac{1}{|\mathbf{r}_1 - \mathbf{r}_2|} \left(\int_{\mathbb{R}^{3N-6}} \int_{\mathbb{R}^{3N}} \prod_{i=1}^{N} |\varphi_i(\mathbf{r}_i)|^2 \prod_{j=1}^{N} |\varphi_j(\mathbf{u}_j)|^2 \right.$$
$$\left. d^{3N} u \, dr_3 \, dr_4 \ldots dr_N \right) dr_1 \, dr_2.$$

Noting that

$$n(\mathbf{r}_1) = N \int_{\mathbb{R}^{3N-3}} \prod_{i=1}^{N} |\varphi_i(\mathbf{r}_i)|^2 \, dr_2 \, dr_3 \ldots dr_N$$

$$= N \int_{\mathbb{R}^3} \int_{\mathbb{R}^{3N-6}} \prod_{\substack{i=1 \\ (i \neq 2)}}^{N} |\varphi_i(\mathbf{r}_i)|^2 |\varphi_2(\mathbf{u}_2)|^2 \, du_2 \, dr_3 \ldots dr_N$$

and, similarly,

$$n(\mathbf{r}_2) = N \int_{\mathbb{R}^3} \int_{\mathbb{R}^{3N-6}} \prod_{\substack{j=1 \\ (j \neq 2)}}^{N} |\varphi_j(\mathbf{u}_j)|^2 |\varphi_2(\mathbf{r}_2)|^2 \, du_1 \, du_3 \ldots du_N,$$

we see that the first term reduces to

$$c \int_{\mathbb{R}^3} \int_{\mathbb{R}^3} \frac{1}{|\mathbf{r}_1 - \mathbf{r}_2|} \frac{n(\mathbf{r}_1)}{N} \frac{n(\mathbf{r}_2)}{N} \, dr_1 \, dr_2.$$

Now treating each term in the sum in the same manner, summing up the remaining terms, and using $\mathbf{r}$ and $\mathbf{r}'$ as the dummy integration variables, we arrive at (15.34) since there are $N(N-1)/2$ terms in $\sum_i^N \sum_{j>i}^N$. $\square$

The approximation (15.34) provides a representation of the electron–electron interaction energy as a functional of the electron density, $n(\mathbf{r}) = \sum_{i=1}^{N} |\varphi_i(\mathbf{r})|^2$, referred to as the *mean-field* approximation. The prevailing practice is to consider the sum over all possible pairwise interactions in the system so that $N(N-1)/2$ is replaced by $N^2/2$ and the coefficient of the double integral reduces to $c/2$, which is independent of N. The potential field

$$v_H(n) = c \int_{\mathbb{R}^3} \frac{n(\mathbf{r}')}{|\mathbf{r} - \mathbf{r}'|} \, dr'$$

is referred to as the *Hartree potential*, and with the summation convention just mentioned, the total electron–electron interaction energy is approximated by the functional

$$\frac{1}{2} \langle v_H(n), n \rangle = \frac{c}{2} \int_{\mathbb{R}^3} \int_{\mathbb{R}^3} \frac{n(\mathbf{r}) n(\mathbf{r}')}{|\mathbf{r} - \mathbf{r}'|} \, dr \, dr'. \tag{15.35}$$

15.4.2 The Hohenberg–Kohn Theorem

A major contribution to quantum mechanics involved the observation that the complete electronic structure of a multielectron system was determined by the electron density $n(\mathbf{r})$. Thus, a single scalar-valued function of position could possibly replace the multipositioned wave function $\psi(\mathbf{r}_1, \mathbf{r}_2, \ldots, \mathbf{r}_N)$ as a quantity that completely determines the dynamical state of the system. A key idea is to presume that an external potential $v = v(\mathbf{r})$ exists that defines the nuclear–electronic energy, a fact embodied in the *Hohenberg–Kohn Theorem*:

Theorem 15.A *The ground state electron density $n(\mathbf{r})$ of a bound system of electrons in an external potential $v(\mathbf{r})$ determines the potential uniquely to within a constant. Moreover, there exists a functional of electron densities that assumes a minimum at the ground state density.*

Proof We first demonstrate that an energy functional exists that assumes a minimum at the ground state electron density. Clearly, there may be many wave functions ψ that can produce a given density n. Let $\mathcal{V}_n$ denote a subspace of such wave functions that generate n. For a dynamical operator $\tilde{Q}$ we define the n-dependent expected value,

$$\langle Q \rangle_n = Q[n] = \inf_{\psi \in \mathcal{V}_n} \left\langle \psi, \tilde{Q}\psi \right\rangle. \tag{15.36}$$

From part (ii) of Lemma 15.1 (recall (15.33)), the action of a potential $v(\mathbf{r})$ on $n(\mathbf{r})$ is given by

$$\int_{\mathbb{R}^3} v(\mathbf{r})n(\mathbf{r}) \, dr = \left\langle \psi, \sum_{i=1}^{N} v(\mathbf{r}_i)\psi \right\rangle.$$

Let $E[n]$ denote the functional

$$E[n] = Q[n] + \int_{\mathbb{R}^3} v(\mathbf{r})n(\mathbf{r}) \, dr.$$

Then, if

$$\tilde{Q} = \sum_{i=1}^{N} \left(-\frac{\hbar^2}{2m}\Delta_{\mathbf{r}_i} + \sum_{j>i} \frac{e^2}{4\pi\epsilon_0 r_{ij}} \right),$$

we have

$$E[n] = \left\langle \psi, (\tilde{Q} + v)\psi \right\rangle,$$

and $\tilde{Q} + v = H$, the Hamiltonian of the system. Thus, if E_0 is the ground state energy, we know from the variational principle of Chapter 13 that

$$E[n] \geq E_0.$$

Let n_0 be the electron density of the ground state and ψ_0 a corresponding wave function. From (15.36), we obtain

$$Q[n_0] \leq \left\langle \psi_0, \tilde{Q}[n_0]\psi_0 \right\rangle.$$

Thus,

$$\left\langle \psi_0, \tilde{Q}[n_0]\psi_0 \right\rangle + \int_{\mathbb{R}_5} v(\mathbf{r})n_0(\mathbf{r}) \, dr = E[n_0]$$

$$\leq \left\langle \psi, \left(\tilde{Q}[n_0] + \sum_{i=1}^N v(\mathbf{r}_i) \right) \psi \right\rangle = E_0,$$

which means that $E[n_0] = E_0$.

We have established that an energy functional $E[n]$ of the electron density exists that assumes a minimum at the ground state density n_0, assuming that for every density n a unique (to within a constant) potential $v(\mathbf{r})$ exists. To show that this uniqueness prevails, we assume, to the contrary, that two potentials v_1 and v_2 exist that correspond to the same ground state density. If H_1 and H_2 are the corresponding Hamiltonians, and ψ_1 and ψ_2 are respectively nondegenerate ground state wave functions for these Hamiltonians, the ground state energies satisfy

$$E_1 = \langle \psi_1, H_1\psi_1 \rangle = \int_{\mathbb{R}^3} n(\mathbf{r}) \, v_1(\mathbf{r}) \, dr + \langle \psi_1, (T_e + V_{ee})\psi_1 \rangle,$$

$$E_2 = \langle \psi_2, H_2\psi_2 \rangle = \int_{\mathbb{R}^3} n(\mathbf{r}) \, v_2(\mathbf{r}) \, dr + \langle \psi_2, (T_e + V_{ee})\psi_2 \rangle.$$

But then

$$E_1 < \langle \psi_2, H_1\psi_2 \rangle = \int_{\mathbb{R}^3} n(\mathbf{r}) \, v_1(\mathbf{r}) \, dr + \langle \psi_2, (T_e + V_{ee})\psi_2 \rangle,$$

$$= E_2 + \int_{\mathbb{R}^3} n(\mathbf{r}) \left(v_1(\mathbf{r}) - v_2(\mathbf{r}) \right) \, dr.$$

Likewise,

$$E_2 < \langle \psi_1, H_2\psi_1 \rangle = E_1 + \int_{\mathbb{R}^3} n(\mathbf{r}) \left(v_2(\mathbf{r}) - v_1(\mathbf{r}) \right) \, dr.$$

Adding the last two inequalities gives

$$E_1 + E_2 < E_2 + E_1,$$

a contradiction. Thus, there cannot exist a potential v_2 unequal to constant $+ v_1$ that can lead to the same density $n(\mathbf{r})$. $\qquad\square$

We remark that the fact that the Hamiltonian is determined uniquely (to within a constant) by the ground state electron density for any system of electrons is referred to as the First Hohenberg–Kohn theorem. The Second Hohenberg–Kohn theorem asserts that the ground state in a particular external potential field can be found by minimizing the corresponding energy functional. We can now use the standard variational method to seek the minimizer of $E[n]$ over the space of admissible trial densities, $\bar{n}$.

15.4.3 The Kohn–Sham Theory

The development of the Kohn–Sham Theory for electronic systems is built on a formulation akin to the Hartree and Hartree–Fock methods in that it begins with systems of noninteracting electrons. But it then extends the theory by adding corrections to the kinetic energy that modifies the individual electron equations so that they provide an exact characterization of the system, provided that an exchange-correlation energy is known.

Thus, we begin with a system of N noninteracting electrons each in an orbital χ_i, defined by the Hamiltonian,

$$H = \sum_{i=1}^{N} h(\mathbf{r}_i), \tag{15.37}$$

where h is defined in (15.13). This operator has an exact eigenfunction that is the determinant constructed from the eigenstates of the one-electron equations,

$$\left(-\frac{\hbar^2}{2m}\Delta_{\mathbf{r}} + v_e(\mathbf{r}) \right)\psi_i = \epsilon_i \psi_i, \tag{15.38}$$

v_e being the potential for the noninteracting system and i may denote both spatial and spin quantum numbers. The total energy of the noninteracting system is

$$E_{\text{non}}[n] = T_{\text{non}}[n] + \int_{\mathbb{R}^3} n(\mathbf{r})v_e(\mathbf{r})\,dr, \tag{15.39}$$

where now [recall (15.30)]

$$T_{\text{non}}[n] = \sum_{i=1}^{N} \left\langle \psi_i, -\frac{\hbar^2}{2m} \Delta_{\mathbf{r}_i} \psi_i \right\rangle, \tag{15.40}$$

$$n(\mathbf{r}) = \sum_{i=1}^{N} \mathbb{P}_i(\mathbf{r}) = \sum_{i=1}^{N} |\psi_i(\mathbf{r})|^2, \tag{15.41}$$

T_{non} being the total kinetic energy of the noninteracting system. The relation (15.41) follows from the assumption that the one-electron orbitals ψ_i are orthonormal: $\langle \psi_i, \psi_j \rangle = \delta_{ij}$. Then

$$\mathbb{P}_i(\mathbf{r}) = \int_{\mathbb{R}^3} \psi_1^* \psi_1 \, d^3 r_1 \int_{\mathbb{R}^3} \psi_2^* \psi_2 \, d^3 r_2 \times \cdots$$

$$\times \int_{\mathbb{R}^3} \psi_i^*(\mathbf{r}) \psi_i(\mathbf{r}) \, d^3 r \int_{\mathbb{R}^3} \psi_{i+1}^* \psi_{i+1} \, d^3 r_{i+1} \times \cdots$$

$$\times \int_{\mathbb{R}^3} \psi_N^* \psi_N \, d^3 r_N$$

$$= 1 \times 1 \times \cdots \times |\psi_i|^2 \times 1 \times \cdots \times 1$$

$$= |\psi_i|^2.$$

We wish to minimize $E_{\text{non}}[n]$ subject to the constraint $\int_{\mathbb{R}^3} n \, dr = N$. We introduce the Lagrangian,

$$\mathcal{L}([\bar{n}], \bar{\mu}) = E_{\text{non}}[\bar{n}] - \bar{\mu} \left(\int_{\mathbb{R}^3} \bar{n} \, dr - N \right),$$

where $\bar{\mu}$ is the Lagrange multiplier. The Gâteaux (variational) derivative of $T_{\text{non}}[n]$ is defined by

$$\langle DT_{\text{non}}[n], \bar{n} \rangle = \lim_{\theta \to 0} \frac{1}{\theta} \left(T_{\text{non}}[n + \theta \bar{n}] - T_{\text{non}}[n] \right).$$

Here $\langle \cdot, \cdot \rangle$ denotes duality pairing on the space of electron densities and not necessarily an L^2-inner product. Thus, at the minimizer n we have

$$\langle DT_{\text{non}}[n] + v_e(\mathbf{r}) - \mu, \bar{n} \rangle = 0 \qquad \forall \bar{n},$$

so that the Lagrange multiplier is

$$\mu = v_e(\mathbf{r}) + DT_{\text{non}}[n].\qquad(15.42)$$

The function μ is called the *chemical potential* of the noninteracting system. Multiplying (15.38) by ψ_i^* and integrating over $\mathbb{R}^3$ (with appropriate summing of spins) leads to the noninteracting energy as $E[n] = \sum_{i=1}^n \epsilon_i$.

The exact energy of the full interacting system is

$$E[n] = T[n] + \int_{\mathbb{R}^3} v(\mathbf{r})n(\mathbf{r}) \, dr + J[n],\qquad(15.43)$$

where

$$J[n] = \left\langle \psi, \sum_{i=1}^N \sum_{j>i}^N U(\mathbf{r}_i, \mathbf{r}_j)\psi \right\rangle.\qquad(15.44)$$

We can now define a correction to the noninteracting kinetic energy as

$$E_{\text{xc}}[n] \equiv T[n] - T_{\text{non}}[n] + J[n] - \frac{c}{2} \int_{\mathbb{R}^3} \int_{\mathbb{R}^3} \frac{n(\mathbf{r})n(\mathbf{r}')}{|\mathbf{r} - \mathbf{r}'|} \, dr \, dr',$$

$$(15.45)$$

where the last term on the right-hand side is recognized as the Hartree approximation of the electron–electron interaction energy given in (15.35). The functional $E_{\text{xc}}[n]$ is called the *exchange correlation energy*. It is reminiscent of the exchange energy K_{ij} of the Hartree–Fock equations, but quite different in meaning and structure.

For an arbitrary density $\bar{n}(\mathbf{r})$, the *exact energy for a system of interacting electrons is now*

$$E[\bar{n}] = T_{\text{non}}[\bar{n}] + \int_{\mathbb{R}^3} v(\mathbf{r})\bar{n}(\mathbf{r}) \, dr$$
$$+ \frac{c}{2} \int_{\mathbb{R}^3} \int_{\mathbb{R}^3} \frac{\bar{n}(\mathbf{r})\bar{n}(\mathbf{r}')}{|\mathbf{r} - \mathbf{r}'|} \, dr \, dr' + E_{\text{xc}}[\bar{n}].\qquad(15.46)$$

The corresponding constrained Hohenberg–Kohn variational problem is

$$E[n] = \inf_{\bar{n}} \left\{ E[\bar{n}] \; ; \; \int_{\mathbb{R}^3} \bar{n}(\mathbf{r}) dr = N \right\}.\qquad(15.47)$$

Setting up once again the Lagrangian,

$$\mathcal{L}([\bar{n}], \bar{\mu}) = E[\bar{n}] - \bar{\mu}\left(\int_{\mathbb{R}^3} \bar{n} \, dr - N\right),$$

where now $E[\bar{n}]$ is given by (15.46), we obtain for the Lagrange multiplier the new chemical potential,

$$\mu^* = v_{\text{eff}}(\mathbf{r}) + DT_{\text{non}}[n(\mathbf{r})], \tag{15.48}$$

where

$$v_{\text{eff}}(\mathbf{r}) = v(\mathbf{r}) + c\int_{\mathbb{R}^3} \frac{n(\mathbf{r}')}{|\mathbf{r} - \mathbf{r}'|} \, dr' + v_{\text{xc}}(\mathbf{r}), \tag{15.49}$$

and

$$\langle DE_{\text{xc}}[n], \bar{n} \rangle \equiv \int_{\mathbb{R}^3} v_{\text{xc}}(\mathbf{r}) \bar{n}(\mathbf{r}) \, dr. \tag{15.50}$$

In (15.48) and (15.50), D denotes the variational or Gâteaux derivative defined earlier.

We now make a fundamental observation: The form of the chemical potential μ^* in (15.48) is the same as the noninteracting chemical potential μ of (15.42) except that the interacting particles are moving in a potential field v_{eff} instead of v_e. Thus, the minimizing density $n(\mathbf{r})$ can be determined by solving the system of single electron equations,

$$\left(-\frac{\hbar^2}{2m}\Delta_{\mathbf{r}} + v_{\text{eff}}(\mathbf{r}_i)\right)\tilde{\varphi}_i(\mathbf{r}) = \hat{\epsilon}_i\tilde{\varphi}_i(\mathbf{r}), \qquad 1 \leq i \leq N,$$

$$n(\mathbf{r}) = \sum_{i=1}^{N} |\tilde{\varphi}_i(\mathbf{r})|^2, \tag{15.51}$$

$$v_{\text{eff}}(\mathbf{r}) = v(\mathbf{r}) + c\int_{\mathbb{R}^3} \frac{n(\mathbf{r}')}{|\mathbf{r} - \mathbf{r}'|} \, dr' + v_{\text{xc}}(\mathbf{r}).$$

These equations are called the *Kohn–Sham equations*.

To obtain the corresponding ground state energy, we multiply (15.51)$_1$ by $\tilde{\varphi}_i^*$, integrate over $\mathbb{R}^3$, use the fact that the $\tilde{\varphi}_i$ are orthonormal ($\int_{\mathbb{R}^3} \tilde{\varphi}_i^2 \, dr = 1$), and sum over N to obtain

$$T_{\text{non}}[n] + \int_{\mathbb{R}^3} v_{\text{eff}}(\mathbf{r}) n(\mathbf{r}) \, dr = \sum_{i=1}^{N} \hat{\epsilon}_i. \tag{15.52}$$

Introducing this result into (15.46) gives

$$E[n] = \sum_{i=1}^{N} \hat{\epsilon}_i + E_{\text{xc}}[n(\mathbf{r})] - \int_{\mathbb{R}^3} v_{\text{xc}}(\mathbf{r}) n(\mathbf{r}) dr - \frac{c}{2} \int_{\mathbb{R}^3} \int_{\mathbb{R}^3} \frac{n(\mathbf{r}) n(\mathbf{r}')}{|\mathbf{r} - \mathbf{r}'|} \, dr dr'.$$
$$\tag{15.53}$$

We observe that by omitting E_{xc} and v_{xc}, the above energy functional and the Kohn–Sham equations reduce to the form of the Hartree–Fock equations. Kohn [32] refers to (15.51) as an "exactification" of the Hartree–Fock equations.

The Kohn–Sham theory assumes that the electron density is "noninteracting v-representable," meaning that there must exist a potential that will produce the same density as the exact wave function.

The key to the Kohn–Sham Density Functional Theory is the exchange-correlation energy $E_{\text{xc}}[n]$, which is, in general, unknown. Many semiempirical expressions for E_{xc} for specific electronic structures have been proposed. One example is the Local Density Approximation, in which $E_{\text{xc}}^{\text{LDA}} = \int_{\mathbb{R}^3} e_{\text{xc}} n \, dr$, and e_{xc} is the exchange correlation density for a uniform electron gas (see Kohn and Sham [30]). For other forms of E_{xc}, see Gill, Adamson, and Pople [19].

Part III
Statistical Mechanics

BASIC CONCEPTS: ENSEMBLES, DISTRIBUTION FUNCTIONS, AND AVERAGES

16.1 Introductory Remarks

Statistical mechanics is a branch of applied physics devoted to characterizing bulk or macroscopic properties of matter on the basis of the dynamical behavior of microscopic constituents, the atoms or molecules that make up material bodies. Among its most important applications is thermodynamics, the macroscopic theory of heat and temperature, but it also provides fundamental results on how microscopic events lead to macroproperties of gases, liquids, and solid materials. Statistical mechanics thus addresses a daunting question: How can the behavior of many trillions of particles in rapid motion (e.g., Avogadro's number of $\sim 10^{23}$ atoms per mole) lead to the bulk events we experience in every-day contact with matter? We will lay out the elements of classical statistical mechanics for systems in equilibrium, a concept that we develop forthwith.

An Introduction to Mathematical Modeling: A Course in Mechanics, First Edition. By J. Tinsley Oden © 2011 John Wiley & Sons, Inc. Published 2011 by John Wiley & Sons, Inc.

Our focus here is on classical statistical mechanics, which is concerned with the equilibrium properties of matter at ordinary temperature levels (not near absolute zero or at extremely high astronomical levels where relativistic effects may be important). Roughly speaking, equilibrium properties are those macroscopic properties of bodies we observe over time intervals enormously longer than those of events at the atomic or molecular level. While equilibrium properties at a given temperature appear to be constant over a period of time (e.g., microseconds or seconds or minutes), we know that at a microscopic level, materials consist of a huge number of atoms or molecules in violent motions that take place over femtoseconds (10^{-15} sec) or nanoseconds (10^{-9} sec). Thus, the macro-time scale is, for all practical purposes, infinite in comparison to the micro-time scale. On the macro scale, what we observe is some sort of average effect of the events at atomic scales. One goal of statistical mechanics is to provide ideas on how such averages can be computed and understood.

In developing a framework to derive average properties, we keep in mind that equilibrium states are generally reached after a transition period which may be very long in comparison to micro-time scales. For example, two bodies brought in contact exchange energy over a period of time, called the relaxation time, until they achieve a state of thermal equilibrium, at which point they share the same temperature. Exactly what this temperature means in the context of statistical mechanics will come out in later developments.

16.2   Hamiltonian Mechanics

Hamiltonian mechanics provides the most natural framework for discussing classical equilibrium statistical mechanics. In this framework, we consider a physical system of N particles in motion relative to a fixed reference frame. The microstate of the system at time t is defined by the instantaneous positions and momenta of the particles making up the system:

$$N \text{ position coordinate vectors:} \quad \mathbf{q}_1, \mathbf{q}_2, \ldots, \mathbf{q}_N,$$

$$N \text{ momentum coordinate vectors:} \quad \mathbf{p}_1, \mathbf{p}_2, \ldots, \mathbf{p}_N.$$

The set of coordinate pairs,

$$\{\mathbf{q}_i(t),\ \mathbf{p}_i(t)\}_{i=1}^{N},$$

is regarded as a point in a $6N$-dimensional space $\mathcal{H}$ called the *phase space*. The notations $\mathbf{q}_i$ and $\mathbf{p}_i$ denote the position vector and the momentum vector of particle i at time t; but in the presence of geometrical constraints on the motion, $\mathbf{q}_i$ will correspond to generalized coordinates, and $i = 1, 2, \ldots, M < N$. We will employ a fixed Cartesian coordinate system, and we will renumber the components of the coordinates as follows: q_1, q_2, q_3 are the Cartesian components of $\mathbf{q}_1, q_4, q_5, q_6$ are the Cartesian components of $\mathbf{q}_2, \ldots$, and $q_{n-2},\ q_{n-1},\ q_n$ are those of $\mathbf{q}_N$, $n = 3N$, so that a point in $\mathcal{H}$ at time t is denoted

$$\{\mathbf{q}_i(t),\ \mathbf{p}_i(t)\}_{i=1}^{N} = (q_1, q_2, \ldots, q_n,\ p_1, p_2, \ldots, p_n) \equiv (q, p).$$

Hereafter, n denotes the number of generalized coordinates of the particles in the system.

The one-parameter family of microstates,

$$\big(q(t), p(t)\big), \quad t \geq 0,$$

is called a *trajectory* of a path in $\mathcal{H}$, and the time rate of change of states along a trajectory is the *phase velocity*

$$\mathbf{v}(t) = \big(\dot{q}(t), \dot{p}(t)\big), \tag{16.1}$$

$$\dot{q}(t) = \frac{dq(t)}{dt} = \left(\frac{dq_1(t)}{dt}, \ldots, \frac{dq_n(t)}{dt}\right),$$

$$\dot{p}(t) = \frac{dp(t)}{dt} = \left(\frac{dp_1(t)}{dt}, \ldots, \frac{dp_n(t)}{dt}\right). \tag{16.2}$$

The system is generally assumed to be confined to a finite volume V in $\mathbb{R}^3$, so that a representative point must remain within a limited region in $\mathcal{H}$.

16.2.1   The Hamiltonian and the Equations of Motion

For each such dynamical system, we assume that there exists a twice-differentiable functional $H : \mathcal{H} \to \mathbb{R}$ such that

$$\dot{q}_i = \frac{\partial H}{\partial p_i} \quad \text{and} \quad \dot{p}_i = -\frac{\partial H}{\partial q_i}, \qquad i = 1, 2, \ldots, n. \qquad (16.3)$$

The functional H is, in fact, our old friend from Part II, the *Hamiltonian* of the system. Equations (16.3) are the *equations of motion*. We will explain why H is the functional dealt with extensively in Part II.

As to why (16.3) define equations of motion, a straightforward connection with these conditions can be made to Lagrangian dynamics. The Lagrangian L of a dynamical system is the difference between the kinetic energy κ and the internal energy U:

$$L(q, \dot{q}) = \kappa - U, \qquad (16.4)$$

and the equations of motion are embodied in the Euler–Lagrange equations,

$$\frac{d}{dt}\left(\frac{\partial L}{\partial \dot{q}_i}\right) - \frac{\partial L}{\partial q_i} = 0, \qquad i = 1, 2, \ldots, n. \qquad (16.5)$$

To provide a certain symmetry in this system, we define $\partial L/\partial \dot{q}_i$ as the generalized momenta p_i and introduce the Legendre transformation,

$$H(q, p) = \sum_{i=1}^{n} \dot{q}_i p_i - L(q, \dot{q}). \qquad (16.6)$$

The function H, of course, is the Hamiltonian, and the equations of motion (16.3) can be derived from (16.6) and (16.5).

The standard example of the Hamiltonian from the Newtonian dynamics of a system of particles of masses m_1, m_2, $m_3 = m_{(1)}$ for particle 1, m_4, m_5, $m_6 = m_{(2)}$ for particle 2, and so forth, is

$$H(q_1, q_2, \ldots, q_n, p_1, p_2, \ldots, p_n) = \sum_{i=1}^{n} \frac{p_i^2}{2m_{(i)}} + V(q_1, q_2, \ldots, q_n)$$

$$= H(q, p),$$

where $V(q_1, q_2, \ldots, q_n)$ is the potential energy of the system (recall (10.13) et seq.).

The Hamiltonian description of the dynamics of the system is clear: for a system with n degrees of freedom, the system moves along trajectories $(q(t), p(t))$, the components of which are solutions of the equations of motion (16.3).

16.3   Phase Functions and Time Averages

For very large systems of particles, such as those consisting of atoms comprising a material body, the macromechanical properties of which are of interest, we know that there is little hope of solving the large system of equations of motion completely. For one thing, information defining the problem is unknown or incomplete (e.g., boundary –and initial– conditions) and, secondly, the enormous size of the system makes a complete solution hopeless. Very fortunately, statistical mechanics allows us to overcome these difficulties by virtue of the fact that only average aspects of the full solution are needed to establish macroscopic equilibrium properties.

To set the stage for computing such averages, we introduce the notion of a phase function to describe any property of a system of N particles that depends on the microstate $(q(t), p(t))$ of the system at time t. Thus, a phase function F is a map from the phase space $\mathcal{H}$ into $\mathbb{R}$ (or into $\mathbb{R}^d$ for vector-valued functions, $d = 2, 3, \ldots$). Examples of phase functions are the Hamiltonian itself, $H(q, p)$, the total kinetic energy $\kappa(q, p) = \sum_{i=1}^{n} p_i^2 / 2m_{(i)}$, etc.

Now any phase function $F(q, p)$ will vary rapidly along any trajectory in phase space compared to a macroscopic time scale. For example, a crystalline metal specimen may be subjected to macroscopic measurements taking seconds, while its period of lattice vibrations may be on the order of nanoseconds (10^{-9} sec). What is measured in a macromeasurement is the time average,

$$F_{\text{avg}}(t, t_0) = \frac{1}{t} \int_{t_0}^{t_0+t} F\big(q(\tau), p(\tau)\big) \, d\tau.$$

Owing to the huge ratio between macro- and micro-time scales, the above average will be essentially independent of t and can be replaced by $\lim F_{\text{avg}}(t, t_0)$ as $t \to \infty$. We denote this average by $\bar{F}$:

$$\bar{F} = \lim_{T \to \infty} \frac{1}{T} \int_0^T F\big(q(\tau), p(\tau)\big) \, d\tau. \qquad (16.7)$$

The actual calculation of the integral in (16.7) is, unfortunately, still hopeless for a macroscopic system because the trajectory $(q(t), p(t))$ involves an enormous number of degrees of freedom. Moreover, the initial conditions are unknown. These difficulties are often overcome for many systems by introducing the so-called *ergodic hypothesis*, discussed in the next section.

16.4   Ensembles, Ensemble Averages, and Ergodic Systems

The concept of an ensemble of systems provides a powerful alternative to the primitive notion of time averaging. The idea is that some physical property of a system under fixed macroscopic conditions, generally observed as a time average of that property over a long-term interval, can instead be realized as the average of a sequence of many short-time observations performed on the same system while its microstate varies over possible states accessible to (or consistent with) the given macrostate. In the ensemble approach, we conceptualize performing many experiments simultaneously on a collection of replicas as opposed to performing experiments sequentially on the system over a long time period. A replica is a manifestation of the given system governed by the same Hamiltonian. The collection of all replicas of a system is called an *ensemble*, a term introduced by J. W. Gibbs. Literature on classical statistical mechanics use other terms to describe replicas, such as (possible) microstates of the system.

Boltzmann is attributed to have introduced the notion of a density or distribution function $\varrho = \varrho(q, p, t)$ which defines the number of representative points in the volume element $dq \, dp$ at time t. The density function describes the distribution of members of an ensemble that are distributed over all possible states at time t. Then $\int_{\mathcal{J}} \varrho \, dq \, dp = n_{\mathcal{J}}$ is the number of members in a subset $\mathcal{J}$ of $\mathcal{H}$. We may normalize ϱ so that $\int_{\mathcal{H}} \varrho \, dq \, dp = 1$. Because of these properties, we can interpret ϱ as a probability distribution function and $\varrho(q, p, t)$ is then the probability that state (q, p) is in the element $dq \, dp$ at time t.

Recognizing that ϱ defines a weight on possible states accessible to the macrosystem, and that ϱ is equivalent to a probability distribution which characterizes the probability that a macrostate (q, p) lies in the volume $dq \, dp$ in $\mathcal{H}$, the *ensemble average* of a phase function $F : \mathcal{H} \to \mathbb{R}$ is precisely the expected value of F corresponding to the density ϱ:

$$\langle F \rangle = \int_{\mathcal{G}} F(q, p) \, \varrho(q, p) \, dq \, dp. \tag{16.8}$$

The ensemble average is also referred to as the phase average of the phase function $F = F(q, p)$.

Most of classical statistical mechanics rests on the hypothesis, usually attributed to Gibbs, that the time average $\bar{F}$ of (16.7) coincides with the ensemble average $\langle F \rangle$ of (16.8):

$$\bar{F} = \langle F \rangle. \tag{16.9}$$

Systems for which (16.9) holds are called *ergodic systems* and the presumption that (16.9) holds is the *ergodic hypothesis*.

An ergodic system is equivalent to one in which the trajectory of a representative particle in $\mathcal{H}$ traverses all points on the energy surface over an infinite time interval $(-\infty, \infty)$. Unfortunately, in 1913 Rosenthal [64] and Plancherel [63] proved that no physical systems could be ergodic. But because the idea leads to powerful conclusions that agree well with actual observations, the idea that "almost all physical systems are essentially ergodic" (Lebowitz [58]) reinforced the notion of *quasi-ergodic* systems: those for which the trajectory of a point can come arbitrarily close to every point on the energy surface (Ehrenfest and Ehrenfest [50]).

It can be shown (e.g., Weiner [66]) that, for ergodic systems, phase averages may be independent of the starting point of the trajectory and that the precise knowledge of the trajectory may not be needed. In particular, let G denote a subspace of the phase space $\mathcal{H}$ and let χ_G denote its characteristic function,

$$\chi_G(q, p) = \begin{cases} 1 & \text{if } (q, p) \in G, \\ 0 & \text{if } (q, p) \notin G. \end{cases} \tag{16.10}$$

Then, for ergodic systems, we have

$$\hat{\chi}_G = \int_{\mathcal{H}} \chi_G \, \varrho \, dq \, dp = \int_G \varrho \, dq \, dp = \lim_{T \to \infty} T^{-1} \int_0^T \chi_G \, d\tau = t_G, \tag{16.11}$$

where t_G is the limiting amount of time the system spends in region G. The result is independent of the starting point of the trajectory. Thus, we have:

> *The integral of ϱ over a subset $G \subset \mathcal{H}$ is the fraction of time the system spends in G.*

Then $\varrho \geq 0$ and $\int_{\mathcal{H}} \varrho \, dq \, dp = 1$, as required. Further, suppose that the time variation of the characteristic function $\chi_G(t)$ fluctuates between 1 and 0 over a time interval. Since only the average $\hat{\chi}_G = \bar{\chi}_G$ is of interest, the order of the 1 and 0 segments can be scrambled without affecting the result. Thus, such averages can be obtained without a precise knowledge of the trajectory of the system. These observations reveal the fundamental importance of ϱ and its explicit form for ergodic systems.

Classical statistical thermodynamics is concerned with the derivation of distribution functions ϱ for systems involving a small number (often just three) of macroscopic properties (such as volume V, number of particles N, energy E, etc.) that represent constraints on the system. These lead to expressions for ϱ for the classical ensembles of statistical thermodynamics. We discuss these examples in Sections 16.6, 16.7, and 16.8.

We remark that, in general, the distribution function ϱ_{ens} for a particular type of ensemble can be written as a product (cf. Allen and Tildesley [45, p. 37] or McQuarrie [59]),

$$\varrho_{ens}(q, p) = \Omega_{ens}^{-1} \, \omega_{ens}(q, p), \tag{16.12}$$

where $\omega_{ens}(q, p)$ is the non-normalized form of $\varrho_{ens}(q, p)$ and Ω_{ens} is the *partition function* for the ensemble; e.g., for the continuous case,

$$\Omega_{ens} = \int_{\mathcal{H}} \omega_{ens}(q, p) \, dq \, dp. \tag{16.13}$$

The quantity Ω_{ens} is a function of the macroscopic properties defining the ensemble, such as (N, V, E) (the number of molecules, the volume of the system, the total energy) or (N, V, T), etc. The function

$$\Pi_{ens} = -\ln \Omega_{ens}. \tag{16.14}$$

is called the *thermodynamic potential* of the ensemble and generally assumes a minimum value at thermodynamic equilibrium for the particular ensemble.

The various ensembles of systems encountered in common applications to equilibrium systems are of special interest. The microcanonical ensemble characterizes a macrostate involving isolated systems characterized by the triple (N, V, E), where N, V, and E, are fixed parameters defining the macrosystem.

- The Microcanonical Ensemble—Isolated System—characterized by (N, V, E).

Of the infinitely many such ensembles and corresponding distribution functions, two of the most important are the following classical ensembles:

- The Canonical Ensemble—Closed Isothermal System—characterized by (N, V, T);

- The Grand Canonical Ensemble—Open Isothermal System—characterized by (μ, V, T).

Again N is the total number of particles in the system, V its volume, E the total energy, T the temperature, and μ the chemical potential.

We discuss isolated systems in the next sections and describe the above three classical ensembles in Sections 16.6, 16.7, and 16.8.

16.5   Statistical Mechanics of Isolated Systems

In this section, we confine ourselves to so-called *isolated systems*, which are in fact one of the main focuses of classical statistical mechanics. A system is isolated if and only if its Hamiltonian depends only on the phase coordinates (the generalized coordinates and momenta) of the system, and is independent of those particles outside the system. The phase space $\mathcal{H}$ thus contains all possible paths of the system consistent with kinematical constraints that may be imposed. A subset of these paths satisfy the equations of motion and are said to be dynamically possible trajectories, and in this subset, the state of the system at one time instant determines its state at later times. Since the equations of motion are first-order ordinary differential equations:

> *Dynamically possible trajectories in an isolated system do not intersect one another, and one and only one dynamically possible trajectory passes through each point in $\mathcal{H}$.*

An important property of isolated systems involving the density function ϱ is given in the following theorem.

Theorem 16.A (Liouville's Theorem) *In any motion of an ensemble of replicas in the phase space of an isolated Hamiltonian system, the distribution function ϱ is constant along any trajectory.*

Proof Let ω denote a volume in the phase space $\mathcal{H}$ and let $\partial\omega$ be the surface enclosing ω. The rate of change of $\varrho\, dq\, dp = \varrho\, d\omega$ is

$$\frac{\partial}{\partial t}\int_\omega \varrho\, d\omega,$$

and the net rate of points flowing out of ω as we run over replicas of the system is

$$\int_{\partial\omega} \varrho\, \mathbf{v}\cdot\mathbf{n}\, d\sigma,$$

where $\mathbf{v} = (\dot{q}, \dot{p})$ is the velocity vector, $\mathbf{n}$ is the unit exterior normal to $\partial\omega$, and $d\sigma$ is a surface element. By the divergence theorem, we have

$$\int_{\partial\omega}(\varrho\mathbf{v})\cdot\mathbf{n}\, d\sigma = \int_\omega \operatorname{div}(\varrho\mathbf{v})\, d\omega,$$

where

$$\operatorname{div}(\varrho\mathbf{v}) = \sum_{i=1}^{2n}\frac{\partial}{\partial y_i}(\varrho v_i) \qquad \big(y_i = q_i\,,\ y_{i+n} = p_i\,,\ v_i = \dot{q}_i\,,\ v_{i+n} = \dot{p}_i\,,$$

$$i = 1, 2, \ldots, n\big)$$

$$= \sum_{i=1}^{n}\left(\frac{\partial\varrho}{\partial q_i}\dot{q}_i + \frac{\partial\varrho}{\partial p_i}\dot{p}_i\right) + \underbrace{\sum_{i=1}^{n}\varrho\left(\frac{\partial^2 H}{\partial q_i\partial p_i} - \frac{\partial^2 H}{\partial p_i\partial q_i}\right)}_{=0}$$

$$= \{\varrho, H\},$$

where

$$\{\varrho, H\} := \sum_{i=1}^{n}\left(\frac{\partial\varrho}{\partial q_i}\frac{\partial H}{\partial p_i} - \frac{\partial\varrho}{\partial p_i}\frac{\partial H}{\partial q_i}\right). \tag{16.15}$$

The notation $\{\cdot, \cdot\}$ is called the *Poisson bracket*, and so the divergence of the "phase flow" $\varrho \mathbf{v}$ in an isolated Hamiltonian system coincides with the Poisson bracket of $\{\varrho, H\}$.

Continuing, we obtain

$$\int_\omega \frac{\partial \varrho}{\partial t}\, d\omega = -\int_\omega \mathrm{div}(\varrho \mathbf{v})\, d\omega,$$

or

$$\int_\omega \left(\frac{\partial \varrho}{\partial t} + \mathrm{div}(\varrho \mathbf{v})\right) d\omega = 0.$$

But this can hold for arbitrary ω only if

$$\frac{d\varrho}{dt} = \frac{\partial \varrho}{\partial t} + \mathrm{div}(\varrho \mathbf{v}) \equiv \frac{\partial \varrho}{\partial t} + \{\varrho, H\} = 0. \qquad (16.16)$$

Thus, ϱ is constant along any trajectory, i.e., any set of points (q, p) satisfying the equations of motion. $\qquad \square$

In general, the systems of interest here are said to be *stationary*, meaning that ϱ does not depend explicitly on time t. Then

$$\left.\frac{\partial \varrho}{\partial t}\right|_{(q,p)} = 0. \qquad (16.17)$$

Condition (16.17) then is viewed as a condition for equilibrium of the system. We can seek distribution functions for such systems, then, that satisfy for a given Hamiltonian

$$\{\varrho, H\} = 0. \qquad (16.18)$$

Corollary 16.1 (to Liouville's Theorem) *The stationary systems described in Liouville's Theorem are volume preserving.*

Proof If $\varrho = $ constant, $d\varrho/dt + \varrho\, \mathrm{div}\, \mathbf{v} = 0$, which implies that $\mathrm{div}\, \mathbf{v} = 0$ (recall (8.2)). $\qquad \square$

The next result is also fundamental.

Theorem 16.B (Constant Energy Surfaces) *An isolated Hamiltonian system follows a trajectory on which the Hamiltonian is a constant E, and E is defined as the total energy of the system.*

Proof The variation in H along a trajectory is

$$\frac{dH}{dt} = \sum_{i=1}^{n} \left(\frac{\partial H}{\partial q_i} \dot{q}_i + \frac{\partial H}{\partial p_i} \dot{p}_i \right) = \sum_{i=1}^{n} \left(\frac{\partial H}{\partial q_i} \frac{\partial H}{\partial p_i} - \frac{\partial H}{\partial p_i} \frac{\partial H}{\partial q_i} \right) = 0.$$

Hence, H = constant. We set this constant equal to E. $\qquad\square$

Thus, when H does not depend explicitly on time, the trajectory of an isolated system with n degrees of freedom is confined to an $(2n - 1)$-dimensional hypersurface $\mathcal{S}(E)$ for which

$$H(q, p) = E. \tag{16.19}$$

Another way to state Liouville's Theorem is to observe that the phase volume of a manifold $\mathcal{J}_0 \subset \mathcal{H}$ is simply $M = \int_{\mathcal{J}_0} dq \, dp$, which is the measure of $\mathcal{J}_0$. The equations of motion map $\mathcal{J}_0$ at $t = 0$ to $\mathcal{J}_t$ at time t. According to the Corollary to Liouville's Theorem:

The measure of $\mathcal{J}_t$ at time t is the same as the measure of $\mathcal{J}_0$.

Then $\mathcal{J}$ is said to be an *invariant manifold* (or subspace) of $\mathcal{H}$.

We note that condition (16.16) can be written

$$\boxed{L\varrho = 0,} \tag{16.20}$$

where L is *Liouville's operator*:

$$L = \frac{\partial}{\partial t} + \frac{\mathbf{p}}{m} \cdot \nabla_q + \mathbf{F} \cdot \nabla_p, \tag{16.21}$$

$\mathbf{p}/m$ is the velocity field, ∇_q is the gradient with respect to the position coordinates $\left(\mathbf{p}/m \cdot \nabla_q = \sum_i \dot{q}_i \, \partial/\partial q_i\right)$, $\mathbf{F}$ is the force field, and ∇_p is the momentum gradient $\left(\mathbf{F} \cdot \nabla_p = \sum_i \dot{p}_i \, \partial/\partial p_i\right)$. If we generalize this setting to nonequilibrium cases and account for possible collisions of particles, (16.20) becomes

$$L\varrho = \left(\frac{\partial \varrho}{\partial t}\right)_{\text{coll.}}, \tag{16.22}$$

where $(\partial \varrho/\partial t)_{\text{coll.}}$ is the time rate of change of ϱ due to particle collisions. Equation (16.22) is called *Boltzmann's equation*, and it is a cornerstone of the kinetic theory of gases. We discuss this equation and its implication in more detail later.

16.6   The Microcanonical Ensemble

In the microcanonical ensemble, one trajectory passes through each point in phase space. It is assumed that:

All microstates are distributed uniformly, with equal frequency, over all possible states consistent with the given macrostate (N, V, E).

Stated in another way that will become more meaningful later in this section, the probability that a microstate selected at random from an ensemble will be found in the state (N, V, E) is the same for all microstates in the system. Classically, this assumption is called *the principle of equal a priori probabilities*.

The microcanonical ensemble corresponds to an isolated system of fixed volume V, fixed number N of particles (e.g., molecules), and fixed energy E. In this case, ϱ is simply a constant on a constant energy surface $\mathcal{S}(E)$ and zero elsewhere in $\mathcal{H}$. But this choice has some technical difficulties (see Weiner [66]). Instead we identify a shell $\Sigma(E, \delta_E)$ of width δ_E such that

$$\Sigma(E, \delta_E) = \left\{ (q, p) \in \mathcal{H}, \ E_0 \leq H(q, p) \leq E_0 + \delta_E \right\}.$$

We set

$$\varrho_\delta(q, p) = \begin{cases} \text{constant} & \text{if } (q, p) \in \Sigma(E, \delta_E), \\ 0 & \text{if otherwise.} \end{cases} \tag{16.23}$$

This satisfies (16.17) and the phase average $\hat{F}$ of any phase function F is obtained as the limit, $\hat{F} = \lim_{\delta_E \to 0} \int_{\mathcal{H}} F \varrho_\delta \, dq \, dp$. The distribution function defined by this limiting process is called the *microcanonical distribution*, and the ensemble is called the *microcanonical ensemble*. It is customary to rewrite the limiting distribution symbolically as simply

$$\boxed{\varrho(q, p) = \Omega^{-1} \delta(H(q, p) - E),} \tag{16.24}$$

where Ω is the normalizing factor called the *microcanonical partition function*, also denoted $\Omega(N, V, E)$.

An additional property follows from this limiting process. Suppose we choose a reference level for the energy E such that $E \geq 0$ and move the origin of the phase space $\mathcal{H}$ so that $H(0) = 0$. Let $V(E)$ denote the volume of the level set $\mathcal{V}_E = \{(q, p) : H(q, p) \leq E\}$. Then the phase volume contained in the hypersphere is defined by

$$V(E) = \int_{\mathcal{H}} h\big(E - H(q, p)\big) \, dq \, dp, \tag{16.25}$$

where h is the Heaviside step function,

$$h(x) = \begin{cases} 1, & x > 0, \\ 0, & x \leq 0. \end{cases}$$

The *Khinchin structure function* $K(E)$ is defined as

$$K(E) = \frac{dV(E)}{dE} = {}^{``}\int_{\mathcal{H}}{}^{"} \delta\big(E - H(q,p)\big) \, dq \, dp, \qquad (16.26)$$

where δ is the Dirac delta distribution (" $\int_{\mathcal{H}}$ " emphasizes that $K(E)$ is a distribution and not the action of an integrable function). Since the calculation of K may be difficult, we use instead its Laplace transform,

$$
\begin{aligned}
\Omega(\beta) &= \int_0^\infty K(E) \, e^{-\beta E} \, dE \\
&= \int_0^\infty {}^{``}\int_{\mathcal{H}}{}^{"} \delta\big(E - H(q,p)\big) \, dq \, dp \; e^{-\beta E} \, dE \\
&= \int_{\mathcal{H}} e^{-\beta H(q,p)} \, dq \, dp, \qquad (16.27)
\end{aligned}
$$

β being a real number. The function $\Omega(\beta)$ will appear again in Section 16.7 as the partition function for the canonical ensemble.

16.6.1 Composite Systems

The physical setting envisioned in studying microcanonical ensembles is one in which an isolated system is in a macrostate in which the volume V and the number of particles N are fixed, and E is the total macroscopic energy resulting from the various microstates (replicas) of the system. We say that the triple (N, V, E) characterizes the macrostate of the system, understanding that E may vary somewhat over a shell $\Sigma(E, \delta_E)$.

Thus, for the microcanonical ensemble, we consider two isolated systems, A and B, in macrostates (N_A, V_A, E_A) and (N_B, V_B, E_B). We bring the two systems in contact through a conducting wall, so that volumes V_A and V_B and the number of particles N_A and N_B remain unchanged, but the energies E_A and E_B of the component system change. The total energy of the combined system is

$$E_{A+B} = E_A + E_B + E_{AB} = \text{constant} = C, \qquad (16.28)$$

where E_{AB} defines an interaction energy when the composite system reaches *equilibrium*, E_A will have a value $\bar{E}_A$, and E_B will have a value

$\bar{E}_B = C - (\overline{E_A + E_{AB}})$. The Hamiltonian of the composite system $A+B$ is defined on the product phase spaces $\mathcal{H}_A \times \mathcal{H}_B$. The Hamiltonian of the combined system is [66]

$$H_{A+B}(y_A, y_B) = H_A(y_A) + H_B(y_B) + H_{AB}(y_A, y_B), \qquad (16.29)$$

where $y_A = (q_A, p_A)$, H_A is the Hamiltonian of system A if A were isolated, $y_B = (q_B, p_B)$, and H_B is the Hamiltonian of B isolated. The term H_{AB} defines interactions between A and B.

Two systems, A and B, are said to be in *weak interaction* if H_{AB} is negligible in the computation of any integral over phase space $\mathcal{H}_{A+B}$. Weiner [66] points out that H_{AB} may be omitted in volume integrals for such systems, but H_{AB} may have significant effect on the trajectories in $A + B$. Clearly, for such systems, we may take for purposes of phase-space averaging,

$$H_{A+B}(y_A, y_B) \cong H_A(y_A) + H_B(y_B). \qquad (16.30)$$

The Khinchin structure function $K_{A+B}(E)$ for such weakly interactive systems is the convolution of the structure functions of the subsystems:

$$K_{A+B}(E) = \int_0^E K_B(E - E_A) K_A(E_A) \, dE_A. \qquad (16.31)$$

This can be established by noting that the volume $V_{A+B}(E)$ of the region $\{(y_A, y_B) \in \mathcal{H}_{A+B} : H_A(y_A) + H_B(y_B) \leq E\}$ is

$$\int_{V_E^A} V_B(E - H(y_A)) \, dy_A = \int_0^E V_B(E - E_A) K_A(E_A) \, dE_A.$$

Differentiating this last integral with respect to E and observing that $V_B(0) = 0$ gives (16.31). See Weiner [66] for full details.

The idea that microstates in an ensemble may or may not correspond to a given macrostate is significant. The number Ω of microstates corresponding to a given macrostate is called its *multiplicity*. It has an important role in classical statistical mechanics and with the notion of equilibrium of a system. The number of microstates can be thought of as the degeneracy of the energy level E in a quantum mechanics context. For large N, this degeneracy will be extremely high, and various energy levels will be so close to one another that they practically vary continuously.

We imagine that a physical system evolves through a growing number of possible microstates until it comes to rest in a macrostate accessible by the largest possible number of microstates. Comparing two macrostates, the one corresponding to the largest number of microstates is the most probable, and the most probable macrostate corresponds to the most probable state. This, in turn, means that the most probable state, where Ω is a maximum, is the state in which the system spends most of its time, and we identify this state with the equilibrium state of the system. The principle of equal probabilities of microstates leads us to the observation that the probability of a macrostate j for an equilibrium system is $P_j = 1/\Omega$, so Ω is the microcanonical partition function $\Omega(N, V, E)$.

For the microcanonical ensemble, the multiplicity of the composite of two systems, $A + B$, is the product,

$$\Omega_{A+B}(E_A, E_B) = \Omega_A(E_A)\,\Omega_B(E_B)$$
$$= \Omega_A(E_A)\,\Omega_B(E - E_A),$$

where $E = E_A + E_B$ = constant. The combined multiplicity can thus be written as a function of the energy of system A. Since it is, by hypothesis, a maximum at the equilibrium state of the combined system, we have

$$\frac{\partial \Omega_{A+B}}{\partial E_A} = \frac{\partial \Omega_A(E_A)}{\partial E_A} \cdot \Omega_B(E_B) + \Omega_A(E_A)\frac{\partial \Omega_B(E_B)}{\partial E_B} \cdot \underbrace{\frac{\partial E_B}{\partial E_A}}_{=-1} = 0.$$

We conclude that the two systems, when combined, reach equilibrium whenever the energies $E_A = \bar{E}_A$ and $E_B = \bar{E}_B$ are such that

$$\frac{\partial \ln \Omega_A(\bar{E}_A)}{\partial E_A} = \frac{\partial \ln \Omega_B(\bar{E}_B)}{\partial E_B}. \tag{16.32}$$

Denoting

$$\frac{\partial \ln \Omega(E)}{\partial E} = \beta, \tag{16.33}$$

we see that at equilibrium the two systems share a common value of β. The quantity β, then, is related to the *thermodynamic temperature* of the two systems. It is customary to write $\beta = 1/kT$, where k is called the Boltzmann constant ($k = 1.3806 \times 10^{-16}$ erg$/K$) and T is the absolute temperature.

Returning to (16.12), (16.13), and (16.24), we observe that for the microcanonical ensemble, the microcanonical partition function can be written symbolically as

$$\Omega = \kappa \int_{\mathcal{H}} \delta(H(q,p) - E)\, dq\, dp, \qquad (16.34)$$

where $\kappa = N! h^{3N}$ (h being Planck's constant) is a factor designed to handle the indistinguishability of the molecules (see Allen and Tildesley [45]). The thermodynamic potential Π_{ens} of (16.14) is, in this case, a constant (the reciprocal of the Boltzmann constant) times the negative of S, the entropy,[1]

$$S = k \ln \Omega. \qquad (16.35)$$

A variational interpretation of this result is fundamental to classical thermodynamics. If we partition all of the available states consistent with the equilibrium macrostate, and subject one or more subsystems to a constraint, $N = N^*$, $V = V^*$, $E = E^*$, then $\Omega(N^*, V^*, E^*) \leq \Omega(N, V, E)$, so $S(N, V, E) \geq S(N^*, V^*, E^*)$, meaning that at the actual unconstrained state, the entropy is a maximum. The inequality $S(N, V, E) \geq S(N^*, V^*, E^*)$ is essentially the second law of thermodynamics. The classical interpretation of this result is that, at thermodynamic equilibrium, the randomness or disorder of the molecular structure is a maximum; thus the greater entropy, the greater the disorder.

---

[1] An equation of the form (16.35) is inscribed on the tombstone of Ludwig Boltzmann and is one of three equations referred to as Boltzmann's equation: Eq. (16.22) is another one.

16.7 The Canonical Ensemble

In the canonical ensemble, one imagines that a physical system with fixed N and V is immersed in an infinite heat reservoir or bath which is at a temperature T. The system is closed, in that it does not exchange particles with its surrounding environment, isothermal, in that T is a constant value of the parameter set attained at equilibrium for the accessible microstates. The microstates will only be a function of the energy of that state; i.e., the numbers of the ensemble can carry out an exchange of energy for given T.

Let us return to the notion of multiplicity of Ω of microstates. In elementary probability, the multiplicity of a set of events is the total number of outcomes that can possibly occur. If a certain event A can possibly occur n_A times, and another type of event B can occur n_B times, the total number of possible events is $n_A n_B$. Next consider an ensemble of $\mathcal{N}$ identical systems sharing a total energy $\mathcal{E}$, and suppose that n_k denotes the number of systems with energy E_k. Then the set of numbers $\{n_k\}$ must be such that

$$\sum_k n_k = \mathcal{N},$$
$$\sum_k n_k E_k = \mathcal{E}. \tag{16.36}$$

One way to view this ensemble of systems is to consider a complex molecular–atomistic system with quantum energy levels associated with various electronic orbitals or electronic shells of atoms in the system or whole molecules. The energies E_k can be associated with the highly degenerate eigenvalues of the Hamiltonian in Schrödinger's equation, or in the net energy of a complex molecule. Thus, there may be many states or systems n_k with the energy level E_k.

The number $\mathcal{N}$ is the number of members in the ensemble: the number of microstates described earlier. Any set of energy numbers $\{n_k\}$ represents a possible realization of the system. Various states (replicas) of the global system can be obtained by reshuffling the numbers n_k and their associated energies E_k to obtain many distinct distributions, so long as the constraints (16.36) are satisfied. The multiplicity of this collection of ensembles is the number of different ways these distributions can occur, and it is given by

$$\Omega = \frac{\mathcal{N}!}{n_1! \, n_2! \cdots} . \tag{16.37}$$

Under the hypothesis that all possible states are equally likely to occur, the frequency with which a given set $\{n_k\}$ of distributions may occur is proportional to Ω; and, from arguments laid down earlier, the most probable one will be that for which Ω is a maximum. The problem of finding the equilibrium state thus reduces to the problem of finding the distribution $\{n_k^*\}$ which maximizes Ω subject to the constraints (16.36). Traditionally, this process is carried out for discrete systems and then extended to continuous distribution functions .

Two simplifications can be made immediately: (1) We work with $\ln \Omega$ instead of Ω to reduce the size of the quantities to be manipulated (and to reduce products to sums) and (2) we invoke Stirling's formula to approximate logarithms of large factorials: $\ln(n!) \approx n \ln n - n$. Thus, we wish to find the maximum of the functional

$$\Upsilon(\mathbf{n}) = \Upsilon(n_1, n_2, \ldots) = \mathcal{N} \ln \mathcal{N} - \sum_k n_k \ln n_k, \tag{16.38}$$

subject to (16.36). We incorporate the constraints using the method of Lagrange multipliers. The Lagrangian is

$$\mathcal{L}(\mathbf{n}, \alpha, \beta) = \Upsilon(\mathbf{n}) + \alpha \left(\mathcal{N} - \sum_k n_k \right) + \beta \left(\mathcal{E} - \sum_k n_k E_k \right), \tag{16.39}$$

where α and β are Lagrange multipliers, and the critical points satisfy

$$\frac{\partial \mathcal{L}(\mathbf{n}^*, \alpha^*, \beta^*)}{\partial n_k} = \frac{\partial \Upsilon(\mathbf{n}^*)}{\partial n_k} - \alpha^* - \beta^* E_k = 0, \qquad \forall k,$$

$$\mathcal{N} - \sum_k n_k^* = 0,$$

$$\mathcal{E} - \sum_k n_k^* E_k = 0.$$

Since

$$\frac{\partial \Upsilon(\mathbf{n}^*)}{\partial n_k} = -\ln n_k^* - 1,$$

we have

$$\ln n_k^* = -(1 + \alpha^*) - \beta^* E_k.$$

Thus,

$$n_k^* = \exp\left(-(1 + \alpha^*) - \beta^* E_k\right) = C \exp(-\beta^* E_k),$$

where C is a constant. The constraint conditions demand that $\sum_k n_k^* = \mathcal{N}$, so

$$C = \frac{1}{\mathcal{N}} \sum_k \exp(-\beta^* E_k), \tag{16.40}$$

and

$$\frac{n_k^*}{\mathcal{N}} = \text{the probability of state } n_k^* = \frac{\exp(-\beta E_k)}{\sum_k \exp(-\beta E_k)},$$

where we have written β for β^* understanding that this is the unique value of the multiplier associated with the constraint $\mathcal{E} = \text{constant} = \sum_k n_k E_k$. Since

$$\mathcal{E} = \sum_k E_k n_k^* = \frac{\mathcal{N} \sum_k E_k \exp(-\beta E_k)}{\sum_k \exp(-\beta E_k)},$$

we have

$$U = \frac{\mathcal{E}}{\mathcal{N}} = \langle E \rangle = \sum_k E_k P_k, \tag{16.41}$$

where P_k is the discrete distribution function,

$$\boxed{P_k = \frac{\exp(-\beta E_k)}{\sum_k \exp(-\beta E_k)},} \tag{16.42}$$

and $\langle E_k \rangle$ is the ensemble average of the energy states. The quantity U is the average energy per system in the ensemble. The distribution P_k is the *discrete Boltzmann distribution* for the ensemble.

Now let us consider continuously distributed systems in which the energy levels E_k are phase functions of a system of ensembles in which the total energy $\mathcal{E}$ is constant. In this case, we replace E_k by a phase function $u = u(q, p)$ which, in fact, coincides with the Hamiltonian for the microstate (q, p). The *Boltzmann distribution function* is then

$$\varrho = C \exp\bigl(-\beta H(q, p)\bigr), \qquad (16.43)$$

where C is chosen so that $\int_{\mathcal{H}} \varrho \, dq \, dp = 1$. Thus, in place of (16.42) we have

$$\varrho(q, p) = \frac{\exp\bigl(-\beta H(q, p)\bigr)}{\int_{\mathcal{H}} \exp\bigl(-\beta H(q, p)\bigr) \, dq \, dp}. \qquad (16.44)$$

The number β, we recall (see 16.33), is related to the temperature T: $\beta = 1/kT$.

It is interesting to note that an argument can be made, justifying (16.44) as a consequence of (16.42), that connects the ensemble average $\bar{Q}$ ($=$ "$\langle Q \rangle$") of an (observable) phase function with the corresponding expected (or expectation) value of a corresponding observable of a quantum system. The following steps in making this connection follow the description of Frenkel and Smit [52]:

- Let Q be an observable in a quantum system with quantum Hamiltonian H and let $\langle Q \rangle$ denote its expected value:

$$\langle Q \rangle = \langle \psi, Q\psi \rangle. \qquad (16.45)$$

Here $\langle \cdot, \cdot \rangle$ denotes the L^2-inner product on $\mathbb{R}^3$ and ψ is the wave function.

- The ensemble average of Q is then

$$\bar{Q} = \frac{\sum_n \exp(-E_n/kT) \, \langle \psi_n, Q\psi_n \rangle}{\sum_n \exp(-E_n/kT)},$$

where the sum is taken over all quantum states that contribute to this statistical average. Since

$$E_n = \langle \psi_n, H\psi_n \rangle,$$

where ψ_n are now the eigenstates corresponding to E_n, and we can write

$$\bar{Q} = \frac{\sum_n \langle \psi_n, \exp(-H/kT)\, Q\psi_n \rangle}{\sum_m \langle \psi_m, \exp(-H/kT)\, \psi_m \rangle} = \frac{\mathrm{Tr}\exp(-H/kT)Q}{\mathrm{Tr}\exp(-H/kT)},$$

where Tr denotes the trace of the operator. The trace, we recall, is independent of the basis used to characterize the space in which the ψ_n reside.

- Since $H = \kappa + U$, we conceptually partition the eigenstates into those corresponding to the kinetic energy κ (the momentum operator eigenstates) and the internal energy U. The momentum eigenstates are eigenfunctions of the operator κ. We next note that

$$\exp(-\beta\kappa)\exp(-\beta U) = \exp\bigl(-\beta(\kappa + U)\bigr) + \mathcal{O}([\kappa, U])$$
$$= \exp\bigl(-\beta H\bigr) + \mathcal{O}([\kappa, U]),$$

where $[\kappa, U]$ is the commutator for operator κ and U, which is of order $\hbar$, Planck's constant. Thus, as $\hbar \to 0$, we obtain

$$\mathrm{Tr}\exp(-\beta H) \cong \mathrm{Tr}\exp(-\beta\kappa)\,\mathrm{Tr}\exp(-\beta U).$$

- Let x_n be a basis for the eigenvectors of the energy operator $\tilde{U}$ and let r_m be a basis for those of $\tilde{\kappa}$. Then

$$\mathrm{Tr}\exp(-\beta H) = \sum_n \sum_m \underbrace{\langle x_n, \exp(-\beta U)x_n \rangle}_{\exp(-\beta U(q))} \underbrace{\langle x_n, r_m \rangle}_{V^{-N}}$$
$$\times \underbrace{\langle r_m, \exp(-\beta\kappa)r_m \rangle}_{\exp(-\beta \sum_{i=1}^N p_i^2/2m)}$$
$$= \frac{1}{h^{dN} N!} \int_{\mathcal{H}} \exp\Bigl(-\beta\bigl(\textstyle\sum_{i=1}^N p_i^2/2m + U(q)\bigr)\Bigr)\,dq\,dp,$$

where V is the volume of the system, N is the number of particles, the sums have been replaced by integrals over $\mathcal{H}$, and d is the dimension of the domain ($d = 1, 2$, or 3).

- Similarly, we derive the limit of $\mathrm{Tr}\exp(-\beta H)Q$.

- Summing up, we have

$$\bar{Q} = \langle Q \rangle = \frac{\int_{\mathcal{H}} \exp(-\beta H(q,p)) \, Q(q,p) \, dq \, dp}{\int_{\mathcal{H}} \exp(-\beta H(q,p)) \, dq \, dp}.$$

Hereafter we will denote the average $\bar{Q}$ by $\langle Q \rangle$ since there should be no confusion with this and the quantum expected value (16.45).

Returning to (16.41), the "internal" energy U of the system is, with the interpretation $u(q,p) = H(q,p)$ (if we take the macroscopic kinetic energy to be zero for a system in equilibrium), the ensemble average of the Hamiltonian:

$$U = \langle H \rangle = \int_{\mathcal{H}} H(q,p) \, \varrho(q,p) \, dq \, dp. \tag{16.46}$$

16.8 The Grand Canonical Ensemble

The grand canonical ensemble is a further generalization of the canonical ensemble in which the system is in a heat bath at a temperature T but is open to an exchange of both energy and particles. The number of particles N, then, can vary, but generally fluctuate about mean values. The exchange of particles is characterized by a potential μ, called the chemical potential. Then either μ or T or both can be varied.

The grand canonical ensemble amounts to a set of canonical ensembles, each in thermodynamic equilibrium, but taking on all possible values of N, and each with the same β. The distribution function is of the form

$$\varrho(q,p) = \Omega(\mu, V, T) \exp(-\beta(H(q,p) - \mu N)), \tag{16.47}$$

where μ is the chemical potential and

$$\Omega(\mu, V, T) = \sum_N \int_{\mathcal{H}} \exp(-\beta(H(q,p) - \mu N)) \, dq \, dp. \tag{16.48}$$

For the grand canonical ensemble, the thermodynamic potential is the negative of the product of $\beta = 1/kT$ and pV, the product of the thermodynamic pressure p and the volume V,

$$pV = kT \ln \Omega(\mu, V, T). \tag{16.49}$$

Here the scenario is that a system A of N_A particles and energy E_A is immersed in a large reservoir B and will exchange both particles and energy with B. After a period of time has elapsed, components A and B of the combined system will attain a common temperature T and a common chemical potential μ. At any time in transition to the equilibrium state, one always has

$$N_B + N_A = N = \text{constant} \quad \text{and} \quad E_A + E_B = E_{A+B} = \text{constant},$$

and since that reservoir is assumed to be much larger than A, only N_A and E_A are of interest. In analogy with (16.36), one argues that at the equilibrium state the probability is proportional to the multiplicity $\Omega_B(N_B, E_B) = \Omega_B(N - N_A, E_{A+B} - E_A)$. The derivation of (16.47) then follows from arguments similar to those leading to (16.42) and (16.44). We leave the detailed derivation as an exercise (see Exercises 7 and 8 in Set III.1). Alternative derivations of (16.47) can be found in most texts on statistical mechanics (e.g., [46, 49, 55, 56, 59, 62, 65, 66]).

The ensembles presented are the examples of classical ensembles of equilibrium statistical mechanics, and many other ensembles could be constructed that are relevant in other applications of the theory. Some of these are discussed in texts on statistical mechanics, such as [48, 59].

16.9   Appendix: A Brief Account of Molecular Dynamics

Molecular dynamics refers to a class of mathematical models of systems of atoms or molecules in which

1. each atom (or molecule) is represented by a material point in $\mathbb{R}^3$ assigned to a point mass,

2. the motion of the system of mass points is determined by Newtonian mechanics, and

3. generally it is assumed that no mass is transferred in or out of the system.

In addition to these basic assumptions, virtually all MD calculations also assume that:

4. electrons are in the ground state (electronic motions are not considered) and a *conservative* force field exists that is a function of positions of atoms or molecules only (thus, in harmony with the Born–Oppenheimer approximation, electrons remain in the ground state when atomic nuclei positions change),

5. the conservative force fields are determined by interatomic or intermolecular potentials V, and long-range interactions are (generally) ignored or "cut-off" radii of influence are used to localize the interaction forces, and

6. boundary conditions are (generally) assumed to be periodic.

Molecular dynamics (MD) is generally used to calculate time averages of thermochemical and thermomechanical properties of physical systems representing gases, liquids, or solids.

16.9.1 Newtonian's Equations of Motion

Thus, in MD we consider a collection of N discrete points in $\mathbb{R}^3$ representing the atom or molecule sites in some bounded domain $\Omega \subset \mathbb{R}^3$, each assigned a point mass m_i and each located relative to a fixed origin **O** by a position vector $\mathbf{r}_i$, $i = 1, 2, \ldots, N$. The motion of each such particle is assumed to be governed by Newton's second law:

$$m_i \frac{d^2 \mathbf{r}_i}{dt^2} = \mathbf{F}_i, \qquad i = 1, 2, \ldots, N, \tag{16.50}$$

where $\mathbf{F}_i$ is the net force acting on particle i representing the interactions of i with other particles in the system. These interatomic forces are always assumed to be conservative; i.e., there exists a potential energy $V = V(\mathbf{r}_1, \mathbf{r}_2, \ldots, \mathbf{r}_N)$ such that

$$\mathbf{F}_i = -\frac{\partial V}{\partial \mathbf{r}_i}, \qquad i = 1, 2, \ldots, N, \tag{16.51}$$

so that finally

$$m_i \frac{d^2 \mathbf{r}_i}{dt^2} + \frac{\partial V}{\partial \mathbf{r}_i} = \mathbf{0}, \qquad i = 1, 2, \ldots, N. \tag{16.52}$$

To (16.52) we must add initial conditions on $\mathbf{r}_i(0)$ and $d\mathbf{r}_i(0)/dt$ and boundary conditions on the boundary $\partial\Omega$ of Ω, these generally chosen to represent a periodic pattern of the motion of the system throughout $\mathbb{R}^3$. The solutions $\{\mathbf{r}_i(t)\}_{i=1}^N$ of (16.52) for $t \in [0, T]$ determine the dynamics of the system. Once these are known, other physical properties of the system such as ensemble averages represented by functionals on the $\mathbf{r}_i$ and $d\mathbf{r}_i/dt$ can be calculated. The characterization of the behavior of the dynamical system must also be invariant under changes in the inertial coordinate system and the time frame of reference.

As in quantum systems, it is also possible to describe the equations of MD in the phase space of position–momentum pairs $(\mathbf{q}, \mathbf{p})$ and in terms of the Hamiltonian of the system,

$$H(\mathbf{q}_1, \mathbf{q}_2, \ldots, \mathbf{q}_N; \mathbf{p}_1, \mathbf{p}_2, \ldots, \mathbf{p}_N) = \sum_{i=1}^N \frac{1}{2m_i} \mathbf{p}_i \cdot \mathbf{p}_i + V(\mathbf{q}_1, \mathbf{q}_2, \ldots, \mathbf{q}_N), \tag{16.53}$$

with $\mathbf{q}_i \equiv \mathbf{r}_i$. Then, instead of (16.52), we have the $2N$ first-order systems, described in (16.3).

16.9.2   Potential Functions

One of the most challenging aspects of MD is the identification of the appropriate potential function V for the atomic–molecular system at hand. The following is often given as a general form of such potentials (returning to the notation $\mathbf{q} = \mathbf{r}$):

$$V(\mathbf{r}_1, \ldots, \mathbf{r}_N) = \sum_{\substack{i=1}}^{N} V_1(\mathbf{r}_i) + \sum_{\substack{i,j=1 \\ j>i}}^{N} V_2(\mathbf{r}_i, \mathbf{r}_j) + \sum_{\substack{i,j,k=1 \\ k>j>i}}^{N} V_3(\mathbf{r}_i, \mathbf{r}_j, \mathbf{r}_k) + \cdots,$$

$$(16.54)$$

where V_s is called the *s-body potential*. The one-body potential,

$$V_1(\mathbf{r}_1) + V_1(\mathbf{r}_2) + \cdots + V_1(\mathbf{r}_N),$$

is the potential energy due to the external force field; the two-body potential,

$$V_2(\mathbf{r}_1, \mathbf{r}_2) + V_2(\mathbf{r}_1, \mathbf{r}_3) + \cdots + V_2(\mathbf{r}_{N-1}, \mathbf{r}_N),$$

represents pairwise interactions of particles; the three-body potential represents three-particle interactions, etc. The *Lennard-Jones* potential,

$$V(\mathbf{r}_i, \mathbf{r}_j) = 4\varepsilon \left((\sigma/r_{ij})^{12} - (\sigma/r_{ij})^{6} \right), \qquad r_{ij} = |\mathbf{r}_i - \mathbf{r}_j|, \quad (16.55)$$

σ being a constant that depends upon the atomic structure of the atom or molecule at positions $\mathbf{r}_i, \mathbf{r}_j$, and ε being the potential energy at the minimum of V, is a well-known example of a 2-body potential.

We remark that in computer simulations employing MD models, approximations of potentials are used to simplify the very large computational problems that can arise. For just pairwise interactions of N particles, for example, the force $\partial V/\partial \mathbf{r}_i$ involves $(N^2 - N)/2$ terms. A typical simplification is to introduce a cut-off radius R around each particle and to include only those interactions with neighboring particles inside that radius. The truncated pairwise potential is then

$$V_{\text{cut-off}}(r) = \begin{cases} f(r)V(r), & r \leq R, \\ 0, & r > R, \end{cases}$$

where $r = r_{ij}$ and f is a smooth cut-off function varying from 1 at $r = 0$ to 0 at $r = R$.

A general form of a potential energy functional for a network of molecules consisting of long molecular chains, such as that illustrated in

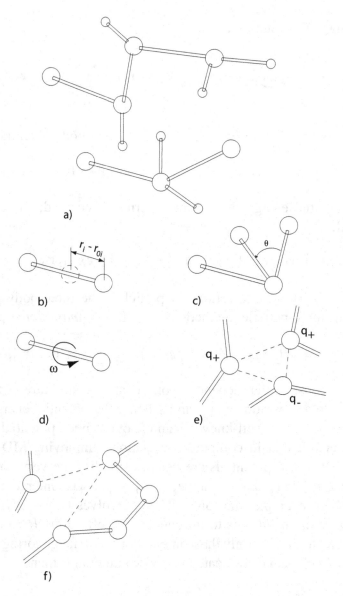

Figure 16.1: (a) A network of molecule chains and side molecules bound by various bonds. (b) Covalent bond stretching. (c) Rotation. (d) Torsion. (e) Nonbound Coulombic interactions. (f) Weak van der Waals bonds.

Fig 16.1, is (cf. [57, p. 166]):

$$V(\mathbf{r}^N) = \sum_{i=1}^{N_{co}} \frac{k_i}{2}(\mathbf{r}_i - \mathbf{r}_{0i})^2 + \sum_{i=1}^{N_a} \frac{\kappa_i}{2}(\theta_i - \theta_{0i})^2$$

$$+ \sum_{i=1}^{N_t} \frac{\kappa_i^t}{2}(1 + \cos(i\omega - \gamma))^2$$

$$+ \sum_{i=1}^{N} \sum_{j=i+1}^{N} \left(4\epsilon_{ij}\left((\sigma_{ij}/r_{ij})^{12} - (\sigma_{ij}/r_{ij})^6\right) + \frac{q_i q_j}{4\pi\epsilon_0 r_{ij}} \right).$$

Here the first term corresponds to stretching of N_{co} covalent bonds with bond stiffness k_i, the second term corresponds to rotational resistance at the junction of a bond length due to rotation of one link with respect to another, term three is due to a torsional rotation about a bond axis, and the final term consists of two parts: nonbonded interactions of a van der Waals bond represented by the 12–6 Lennard-Jones potential, and nonbonded Coulombic interactions between atoms or molecules with charges q_i and q_j. The nonbonded interactions may result from pairs of atoms or molecules that are in different nearby molecules or that are in the same molecule but are separated by at least three bond lengths. The parameters k_i, κ_i, κ_i^t, ϵ_{ij}, γ, and σ_{ij} must be supplied or computed from separate considerations, and $r_{ij} = |\mathbf{r}_i - \mathbf{r}_j|$, etc.

16.9.3 Numerical Solution of the Dynamical System

Once the intermolecular potentials are in hand, we resort to numerical schemes to integrate the equations of motion of the system. One of the most popular numerical schemes for this purpose is the *Verlet Algorithm*.

Assuming that the various fields are analytic in time, we write

$$\mathbf{r}(t + \Delta t) = \mathbf{r}(t) + \Delta t \, \mathbf{v}(t) + \frac{\Delta t^2}{2}\mathbf{a}(t) + \frac{\Delta t^3}{6}\mathbf{b}(t) + \frac{\Delta t^4}{24}\mathbf{c}(t) + \cdots ,$$

$$\mathbf{v}(t + \Delta t) = \mathbf{v}(t) + \Delta t \, \mathbf{a}(t) + \frac{\Delta t^2}{2}\mathbf{b}(t) + \frac{\Delta t^3}{6}\mathbf{c}(t) + \cdots ,$$

$$\mathbf{a}(t + \Delta t) = \mathbf{a}(t) + \Delta t \, \mathbf{b}(t) + \frac{\Delta t^2}{2}\mathbf{c}(t) + \cdots ,$$

$$\mathbf{b}(t + \Delta t) = \mathbf{b}(t) + \Delta t \, \mathbf{c}(t) + \cdots .$$

Truncating:

$$\mathbf{r}(t + \Delta t) = \mathbf{r}(t) + \Delta t \, \mathbf{v}(t) + \frac{1}{2}\Delta t^2 \mathbf{a}(t),$$

$$\mathbf{r}(t - \Delta t) = \mathbf{r}(t) - \Delta t \, \mathbf{v}(t) + \frac{1}{2}\Delta t^2 \mathbf{a}(t).$$

Upon adding these expressions, we eliminate the velocity and obtain

$$\mathbf{r}(t + \Delta t) = 2\mathbf{r}(t) - \mathbf{r}(t - \Delta t) + \Delta t^2 \mathbf{a}(t). \tag{16.56}$$

Velocities can be computed using the central difference approximation,

$$\mathbf{v}(t) = \frac{1}{2\Delta t}\left(\mathbf{r}(t + \Delta t) - \mathbf{r}(t - \Delta t)\right). \tag{16.57}$$

Accelerations are given by

$$\mathbf{a}(t) = \mathbf{F}(t)/m. \tag{16.58}$$

This algorithm needs "starting values": given $\mathbf{r}(0)$ and $\mathbf{a}(0)$, we can estimate $\mathbf{r}(-\Delta t) = \mathbf{r}(0) + \frac{1}{2}\Delta t^2 \mathbf{a}(0) - \Delta t \mathbf{v}(0)$, and use $\mathbf{v}(0) \approx -\Delta t \mathbf{a}(\Delta t)$, for instance.

Once trajectories $(\mathbf{r}_1(t), \ldots, \mathbf{r}_N(t); \mathbf{p}_1(t), \ldots, \mathbf{p}_N(t))$, for $t = 0$, $\Delta t, \ldots, n\Delta t$ are known, we compute macroscale properties of the system by time averaging over some finite time interval $[0, \tau]$. Thus, if

$F = F\big(\mathbf{r}(t), \mathbf{p}(t)\big) \equiv F\big(\mathbf{q}(t), \mathbf{p}(t)\big)$ is some function of the molecular–atomic coordinate pairs, we compute

$$\bar{F} \approx \frac{1}{\tau} \int_0^\tau F\big(\mathbf{q}(t), \mathbf{p}(t)\big)\, dt \approx \frac{1}{\tau} \sum_{i=1}^n F\big(\mathbf{q}(i\Delta t), \mathbf{p}(i\Delta t)\big)\, \Delta t.$$

CHAPTER 17

STATISTICAL MECHANICS BASIS OF CLASSICAL THERMODYNAMICS

17.1 Introductory Remarks

Classical thermodynamics is the study of heat and temperature and the exchange of energy occurring on a *macroscopic scale* in systems in equilibrium or subjected to quasi-static processes. As noted in the preceding chapter, the macroscopic behavior of physical systems is that which we generally observe in nature or in physical experiments involving events we can interpret with our usual senses or with instruments that measure the gross effects of microscopic events. In this chapter we will establish the connection between the statistical mechanics of large systems of particles, atoms, and molecules and the phase-averaged macroscopic equations of mechanics and thermodynamics that we developed in Part I.

We remarked in the preceding chapter that events that occur in the physical systems of interest take place over at least two very different time scales: the micro-time scale in which events at the molecular (microscopic) level take place very rapidly, sometimes at a frequency of

An Introduction to Mathematical Modeling: A Course in Mechanics, First Edition. By J. Tinsley Oden
© 2011 John Wiley & Sons, Inc. Published 2011 by John Wiley & Sons, Inc.

femtoseconds (10^{-15} sec) or nanoseconds (10^{-9} sec), while the events we observe at a macroscale take place in a very slow time scale of microseconds, seconds, minutes, or even hours or longer. We describe these events at the macro-time scale as *quasi-static processes*. They are processes which take a material body, or more precisely material points in a body, through a continuous sequence of states of thermodynamic equilibrium, each governed by the statistical mechanics of equilibrium of the type discussed earlier. Thus, the macro-time variable is a label, a real parameter, identifying one equilibrium state after another, in an arbitrarily slow macro-time scale. The inertia effects in such processes (those depending on the mass of the system) are sometimes negligible, and, therefore, the macroscopic kinetic energy is often assumed to be zero or to be computable from independent considerations such as the quasi-static fluctuations of an ideal gas.

Among the goals of this chapter is to complete in an introductory way the connection between statistical mechanics ensemble averaging, covered up to this point, and the mechanics of continua, covered in Part I. We explore the role of macro constraints on quasi-static processes, following Weiner [66], and we present corresponding Gibbs relations. Brief accounts of Monte Carlo and Metropolis methods are given as means to implement these theories in specific applications. We conclude the study with a brief account of a topic in nonequilibrium statistical mechanics: the study of Boltzmann's equations governing the kinetic theory of gases. This provides a classical connection between statistical mechanics models and continuum gas and fluid dynamics discussed in Part I.

17.2 Energy and the First Law of Thermodynamics

In classical physics, the primitive notion of energy is that it is an intrinsic property of a physical system that defines the capacity of the system, particularly the forces in the system, to do mechanical work or to heat or cool the system. Work, of course, is force times distance, $\int F \, dq$, while heat is a measure of the exchange of energy between two or more systems due to a difference in temperature. Energy is analogous to a potential; when it changes over an instant of time, the change in energy is converted into work or heat.

The total energy in a macrosystem is the sum

$$E = \kappa + U,$$

where κ is the kinetic energy and U is the internal energy of the system. Kinetic energy characterizes the capacity of the system to do work by virtue of its motion—i.e., the work of the inertial forces, mass × acceleration, times displacement per unit time. The internal energy is the capacity of the internal mechanical forces to do work. If $\dot{W}$ denotes the rate at which forces perform work, $\mathcal{P}$ the mechanical power (recall (4.16)), and $\dot{Q}$ the rate of heating or cooling (called the *heating*), then

$$\frac{d}{dt}(\kappa + U) = \dot{W} + \dot{Q}. \tag{17.1}$$

We recognize this relation as the *first law of thermodynamics* (with $\mathcal{P} = \dot{W}$ covered in Part I (recall (5.3))).

For quasi-static thermodynamic processes, $\dot{\kappa}$ is zero or negligible with respect to $\dot{U}$, and slow changes in energy, from one macro-equilibrium state to another occur, so that (17.1) collapses to

$$\dot{U} = \dot{W} + \dot{Q}. \tag{17.2}$$

We emphasize that the superposed dots ($\dot{\ }$) represent time rates relative to the slow macroscopic time scale. We will next reexamine the first law from the context of statistical mechanics.

17.3 Statistical Mechanics Interpretation of the Rate of Work in Quasi-Static Processes

Returning now to the statistical mechanics characterization of quasi-static processes, we know that the physical events of interest can be viewed as occurring at two (or more) scales: the microscale, embodied in the phase space coordinate–momentum pairs (q, p) discussed previously, and macroscale events representing the external constraints on the system imposed by its environment. We visualize a system corresponding to a material body (a solid, fluid, or even a gas), and at a point we prescribe kinematic variables, denoted $G_1, G_2, \ldots, G_K$, or simply G for simplicity. Here G_1 may represent a local volume change ν at a point or a component of strain $G_1 = E_{11}$, $G_2 = E_{12}, \ldots$, or a component of the deformation tensor $G_1 = C_{33}$, or the deformation gradient, $G_i = F_{ij}, j = 1, 2$, or 3. The G's are referred to as *controllable kinematic variables*.

The idea of writing the Hamiltonian as a function of the microstates (q, p) and the kinematic variables G provides a powerful connection between macro- and micro-states and is presented in Weiner [66] and elsewhere in the literature on statistical thermodynamics. Thus, we now write for the Hamiltonian,

$$
\begin{aligned}
H &= H\Big((q_1, q_2, \ldots, q_n, p_1, p_2, \ldots, p_n)\,;\, G_1(t), G_2(t), \ldots, G_K(t)\Big) \\
&= H\big(q, p; G(t)\big), \qquad G(t) = \big(G_1(t), G_2(t), \ldots, G_K(t)\big), \quad (17.3)
\end{aligned}
$$

where t is the macroscale time parameter for quasi-static processes.

We continue to invoke the fundamental ergodic hypothesis of classical statistical mechanics: For any observable macroscale property of a physical system, there is a phase function whose phase average, or ensemble average, is precisely the property observed of the system. Thus, we presume the existence of a distribution function ϱ, as in the preceding chapter, that depends upon the microstate pairs (q, p). For the canonical ensemble, the distribution can be written in the form

$$
\varrho(q, p; G, \theta) = \exp\left(\frac{A(G, \theta) - H(q, p; G)}{\theta}\right), \qquad (17.4)
$$

where $A = A(G, \theta)$ is determined by the normalization condition $\int_{\mathcal{H}} \varrho \, dq \, dp = 1$, and θ is a parameter characterizing the empirical temperature (generally, $\theta = \beta^{-1} = kT$).

Returning to (17.3), we note that the time rate of change of the Hamiltonian H due to changes in the kinematic controllable parameters

G, keeping (q, p) constant, is

$$\frac{dH}{dt}\bigg|_{(q,p)} = \sum_{k=1}^{K} \frac{\partial H}{\partial G_k} \frac{dG_k}{dt}$$

$$= \sum_{k=1}^{K} f_k \, \dot{G}_k, \qquad (17.5)$$

where the f_k are the *micromechanical generalized forces* corresponding to G_k:

$$f_k(q, p; G) = \frac{\partial H(q, p; G)}{\partial G_k}, \qquad k = 1, 2, \dots, K. \qquad (17.6)$$

$\dot{G}_k$ is the slow (macrotime) rate of change of G_k. The forces f_k are phase functions depending on the macro constraints embodied in the kinematic variables G. Thus, if $G = G_1 = v$, a volume change on the microlevel, we obtain

$$f(q, p; v) = \frac{\partial H(q, p; v)}{\partial v} = -p^{\mathrm{m}}(q, p; v),$$

and p^{m} is a phase function such that $\langle p^{\mathrm{m}} \rangle = p =$ the macroscale pressure. If G corresponds to the values of the strain tensor E at a point (e.g., $G_1 = E_{11}, G_2 = E_{12}, \dots, G_6 = E_{33}$), then we write

$$f_{ij}(q, p; E) = \frac{\partial H(q, p; E)}{\partial E_{ij}} = S_{ij}^{\mathrm{m}}(q, p; E), \qquad 1 \le i, j \le 3,$$

where $S_{ij}^{\mathrm{m}}(q, p; E)$ is the phase function for the stress tensor S (the second Piola–Kirchhoff stress) such that $\langle S_{ij}^{\mathrm{m}} \rangle = S_{ij}$. In these examples, we take for granted that a distribution function ϱ is known for the system and that $\langle f \rangle = \int_{\mathcal{H}} f \varrho \, dq \, dp$ for any phase function f.

The *macroscopic generalized forces* F_k are defined as the phase averages of the forces f_k:

$$F_k(G, \theta) = \langle f_k \rangle$$

$$= \int_{\mathcal{H}} \varrho(q, p; G, \theta) \, f_k(q, p; G) \, dq \, dp. \qquad (17.7)$$

The total macroscopic rate of work (i.e., the mechanical power) is the product of the generalized forces F_k and the rate of change of kinematical parameters G_k:

$$\dot{W} = \sum_{k=1}^{K} F_k \, \dot{G}_k. \tag{17.8}$$

Thus, from (17.5), we have

$$\langle \dot{H} \rangle = \left\langle \left. \frac{dH}{dt} \right|_{(q,p)} \right\rangle = \int_{\mathcal{H}} \varrho \sum_{k=1}^{K} f_k \, \dot{G}_k \, dq \, dp$$

$$= \sum_{k=1}^{K} \dot{G}_k \int_{\mathcal{H}} \varrho f_k \, dq \, dp$$

$$= \sum_{k=1}^{K} \dot{G}_k \, F_k.$$

So, from (17.8), we obtain

$$\boxed{\langle \dot{H} \rangle = \dot{W}.} \tag{17.9}$$

We note that this result is independent of the particular form of ϱ.

17.4 Statistical Mechanics Interpretation of the First Law of Thermodynamics

Continuing to consider quasi-static processes on systems for which a distribution ϱ of the form (17.4) of the canonical ensemble is prescribed, we presume that there exists a phase function $u = u(q, p)$ that defines the microscale internal energy, $\langle u \rangle = U$. But for such systems, we know that

$$u = H(q, p; G, \theta), \tag{17.10}$$

so that

$$U = U(G, \theta) = \int_{\mathcal{H}} \varrho(q, p; G, \theta) \, H(q, p; G, \theta) \, dq \, dp, \tag{17.11}$$

which establishes that

$$U = \langle H \rangle. \tag{17.12}$$

The time rate of change of U is then

$$\dot{U} = \frac{d}{dt}\langle H \rangle = \frac{d}{dt}\int_{\mathcal{H}} \varrho H \, dq \, dp$$

$$= \int_{\mathcal{H}} \dot{H}\varrho \, dq \, dp + \int_{\mathcal{H}} H\dot{\varrho} \, dq \, dp$$

$$= \langle \dot{H} \rangle + \int_{\mathcal{H}} H\dot{\varrho} \, dq \, dp. \tag{17.13}$$

We emphasize that these time derivatives are in the slow macrotime regime. Since $H = H(\varrho, q; G(t), \theta)$, we have

$$\dot{H} = \sum_{k=1}^{K} \frac{\partial H}{\partial G_k}\dot{G}_k = \sum_{k=1}^{K} f_k\,\dot{G}_k$$

by (17.6) and (recall (17.9))

$$\langle \dot{H} \rangle = \int_{\mathcal{H}} \varrho \sum_{k=1}^{K} f_k\,\dot{G}_k \, dq \, dp = \dot{W},$$

so we conclude that

$$\dot{Q} = \dot{U} - \dot{W} = \int_{\mathcal{H}} H\dot{\varrho} \, dq \, dp, \tag{17.14}$$

with $\dot{\varrho}$ understood to be defined as

$$\dot{\rho} = \sum_{k=1}^{K} \frac{\partial \varrho}{\partial G_k}\dot{G}_k.$$

Equation (17.14), then, is the statistical mechanics version of the first law of thermodynamics for quasi-static processes.

17.4.1 Statistical Interpretation of $\dot{Q}$

Let us consider systems for which the distribution function is of the form

$$\varrho = e^{\eta}, \tag{17.15}$$

where, for example, for the cases in which (17.4) holds, we obtain

$$\eta(q, p; G(t), \theta) = \frac{A(G(t), \theta) - H(q, p; G(t), \theta)}{\theta}, \tag{17.16}$$

and $\theta = kT$, T being the absolute temperature. Thus,

$$\dot{\varrho} = \dot{\eta} e^{\eta}. \tag{17.17}$$

We establish some straightforward identities in the following lemma:

Lemma 17.1 *Let* (17.15) *and* (17.16) *hold. If A is chosen so that* $\int_{\mathcal{H}} \varrho \, dq \, dp = 1$, *then*

$$(i) \qquad \int_{\mathcal{H}} \dot{\eta} e^{\eta} \, dq \, dp = 0, \tag{17.18}$$

$$(ii) \qquad \int_{\mathcal{H}} A \dot{\varrho} \, dq \, dp = 0, \tag{17.19}$$

and moreover,

$$(iii) \qquad \overline{\langle \dot{\eta} \rangle} = \int_{\mathcal{H}} \eta \dot{\eta} e^{\eta} \, dq \, dp. \tag{17.20}$$

Proof (i) Clearly,

$$\frac{d}{dt} \int_{\mathcal{H}} \varrho \, dq \, dp = \frac{d}{dt}(1) = 0 = \int_{\mathcal{H}} \dot{\varrho} \, dq \, dp = \int_{\mathcal{H}} \dot{\eta} e^{\eta} \, dq \, dp,$$

as asserted.

(ii) This follows because $A = A(G(t), \theta)$ is independent of (q, p).

(iii) We have

$$\dot{\overline{\langle \eta \rangle}} = \frac{d}{dt} \int_{\mathcal{H}} \varrho \eta \, dq \, dp = \int_{\mathcal{H}} \dot{\varrho} \eta \, dq \, dp + \int_{\mathcal{H}} \varrho \dot{\eta} \, dq \, dp.$$

The second integral is zero by (17.18), and $\dot{\varrho}$ is given by (17.17). Hence, (17.20) holds. □

Summing up, we have

$$\dot{Q} = \int_{\mathcal{H}} H \dot{\varrho} \, dq \, dp = \int_{\mathcal{H}} H \dot{\eta} e^{\eta} \, dq \, dp$$

$$= \int_{\mathcal{H}} (H - A) \dot{\eta} e^{\eta} \, dq \, dp \qquad \text{by (17.19)}$$

$$= -\theta \int_{\mathcal{H}} \eta \dot{\eta} e^{\eta} \, dq \, dp.$$

The last integral is recognized as $\dot{\overline{\langle \eta \rangle}}$; see (17.20). Thus,

$$\boxed{\dot{Q} = -\theta \, \dot{\overline{\langle \eta \rangle}}.} \qquad (17.21)$$

Introducing (17.9) and (17.21) into (17.2), we see that the statistical mechanics characterization of the first law of thermodynamics for quasi-static processes is

$$\boxed{\dot{U} = \langle \dot{H} \rangle - \theta \, \dot{\overline{\langle \eta \rangle}}.} \qquad (17.22)$$

17.5 Entropy and the Partition Function

The empirical temperature θ in (17.21) is, as was noted earlier, proportional to the absolute temperature T, the constant of proportionally k being Boltzmann's constant: $\theta = kT$. According to (17.21), the amount of heating $\dot{Q}$ per unit temperature results in a change in a state property S of the system we refer to as its entropy. Thus,

$$\frac{1}{T} \dot{Q} = \dot{S},$$

or, in view of (17.21), we have

$$S(G,T) = -k \langle \eta \rangle. \qquad (17.23)$$

Thus, S is a macroscale property of the system that, for quasi-state processes, depends on G and T and which is proportional to the phase average of the phase function $\eta = \eta(q,p;G,\theta)$ defined in (17.16). Since $\langle \eta \rangle = \int_{\mathcal{H}} \varrho \eta \, dq \, dp$, we have an immediate relationship between the entropy and the distribution function ϱ:

$$S = -k \int_{\mathcal{H}} \varrho \ln \varrho \, dq \, dp. \qquad (17.24)$$

Returning to (17.16), writing $\eta = \eta(G,T)$ since $\theta = kT$, we take the phase average of both sides to obtain

$$kT \langle \eta \rangle = A(G,T) - \langle H \rangle (G,T)$$
$$= A(G,T) - U(G,T),$$

or

$$A = U - ST, \qquad (17.25)$$

where we have omitted writing the dependence of $A, U,$ and S on (G,T). We recognize A as the Helmholtz free energy of the system (recall the definition of the Helmholtz free energy per unit mass, given in (7.13)).

We recall that A appeared in the definition of η so that $\varrho = e^{\eta}$ would satisfy the normalization condition, $\int_{\mathcal{H}} \varrho \, dq \, dp = 1$. Thus,

$$e^{-A/kT} = \int_{\mathcal{H}} e^{-H/kT} \, dq \, dp,$$

or

$$A(G,T) = -kT \ln \Omega(G,T), \qquad (17.26)$$

where

$$\Omega(G,T) = \int_{\mathcal{H}} e^{-H(q,p;G,T)/kT} \, dq \, dp. \qquad (17.27)$$

We recognize that $\Omega(G, T)$ is the *partition function* of the system derived earlier (recall (16.27) or (16.44)). It characterizes the microscopic nature of the system for given (G, T), and its relationship to the Helmholtz free energy, in (17.26), provides the connection between the microscopic or molecular nature of the system and its macroscopic properties.

17.6 Conjugate Hamiltonians

The conjugate J^* of a convex functional $J : V \to \mathbb{R}$, V being a normed space, is defined by

$$J^*(\sigma) = \sup_{v \in V} \left\{ \langle \sigma, v \rangle - J(v) \right\}, \tag{17.28}$$

where $\langle \cdot, \cdot \rangle$ denotes duality pairing on $V' \times V$, V' being the dual space of V. The operation (17.28) is called a *Legendre transformation*, and $J^* : V' \to \mathbb{R}$ is the *conjugate functional* of J (see, e.g., [51, 61, 66] for details). Even when J is not convex, we can define J^* formally as $\sup_{v \in V} \left\{ \langle \sigma, v \rangle - J(v) \right\}$.

Turning to the Hamiltonian H, we write $H = H(q, p; G, T)$, replacing θ with $T = \theta/k = $ constant to conform to more conventional notation. In this case, we shall assume for simplicity that H is convex with respect to G, so that the derivatives with (q, p) and T held fixed are invertible functions. Thus, denoting

$$\boxed{ f_k = \left. \frac{\partial H}{\partial G_k} \right|_{(q,p)} = f_k(q, p; G, T), k = 1, 2, \ldots, K, } \tag{17.29}$$

we compute the inverses,

$$G_k = G_k(q, p; f, T), k = 1, 2, \ldots, K, \tag{17.30}$$

with $f = (f_1, f_2, \ldots, f_K)$. We now define the *conjugate Hamiltonian* H^* through the transformation

$$\boxed{ H^*(q, p; f, T) = H(q, p; G, T) - \sum_{k=1}^{K} f_k \, G_k, } \tag{17.31}$$

Then,

$$
\frac{dH^*}{dt} = \sum_{k=1}^{K} \frac{\partial H^*}{\partial f_k}\bigg|_{(q,p)} \dot{f}_k
$$

$$
= \sum_{k=1}^{K} \left(\frac{\partial H}{\partial G_k} \dot{G}_k - \dot{f}_k \, G_k - \dot{f}_k \, G_k \right),
$$

from which we conclude that

$$
G_k = -\frac{\partial H^*}{\partial f_k}\bigg|_{(q,p)}, \quad k = 1, 2, \ldots, K. \tag{17.32}
$$

The conjugate Hamiltonian provides a dynamical description of the same system described by the Hamiltonian with kinematical central variables G, but with new central parameters F describing the generalized mechanical forces F_k acting on the system. The functionals H and H^* determine the same dynamics for the system under consideration.

When we write the distribution function $\varrho = e^{\eta}$ of the system in terms of the conjugate Hamiltonian, we use the notation

$$
\varrho^*(q, p; F, T) = \exp\left(\frac{A^*(F, T) - H^*(q, p; F, T)}{kT} \right). \tag{17.33}
$$

Then the conjugate Helmholtz free energy is

$$
A^*(F, T) = -k \ln \Omega^*(F, T), \tag{17.34}
$$

where Ω^* is the conjugate partition function,

$$
\Omega^*(F, T) = \int_{\mathcal{H}} \exp\left(\frac{-H^*(q, p; F, T)}{kT} \right) dq \, dp. \tag{17.35}
$$

Various other thermodynamic quantities can be written in terms of H^*, ϱ^*, Ω^*, etc. For example,

$$
U(F, T) = \int_{\mathcal{H}} H^*(q, p; F, T) \, \varrho^*(q, p; F, T) \, dq \, dp, \tag{17.36}
$$

$$
S(F, T) = -k \int_{\mathcal{H}} \varrho^*(q, p; F, T) \ln \varrho^*(q, p; F, T) \, dq \, dp. \tag{17.37}
$$

While the internal energy U satisfies

$$\dot{U} = \sum_{k=1}^{K} F_k \dot{G}_k + \dot{Q},$$

the analogous dual functional,

$$\dot{h} = \sum_{k=1}^{K} G_k \dot{F}_k + \dot{Q}, \tag{17.38}$$

is the rate of change of the *enthalpy*, defined by

$$h = U - \sum_{k=1}^{K} F_k G_k. \tag{17.39}$$

17.7 The Gibbs Relations

An array of fundamental thermodynamic quantities and relations, which we refer to as the Gibbs relations, follow easily from the relationships derived in the previous sections, and their relation to the micro properties of the systems is again reflected by the partition functions Ω and Ω^*. We list them as follows:

- **Internal Energy**

$$U = U(G, T) = -\frac{\partial \ln \Omega}{\partial \beta} = -\frac{1}{\Omega} \frac{\partial \Omega}{\partial \beta}. \tag{17.40}$$

- **Entropy**

$$S(G, T) = k\left(\ln \Omega - \beta \frac{\partial \Omega}{\partial \beta} \right), \tag{17.41}$$

$$S(F, T) = k\left(\ln \Omega^* - \beta \frac{\partial \Omega^*}{\partial \beta} \right). \tag{17.42}$$

- **Helmholtz Free Energy**

$$A(G, T) = -kT \ln \Omega. \tag{17.43}$$

- **Enthalpy**

$$h(F, T) = -\frac{\partial \Omega^*}{\partial \beta}. \tag{17.44}$$

- **Gibbs Free Energy**

$$\mathcal{G}(F, T) = U^*(F, T) - TS^*(F, T) = -kT \ln \Omega^*. \tag{17.45}$$

These relations apply to macroscopic systems characterized by canonical ensembles.

17.8 Monte Carlo and Metropolis Methods

The Gibbs relations, (17.40)–(17.45), and their various alternatives suggest that fundamental macroscopic properties of the most important thermodynamic systems can be defined by the partition function Ω or its conjugate Ω^*. Determining Ω, however, can be a daunting task as it involves the integration of a phase function over the entire $6N$-dimensional phase space $\mathcal{H}$ (recall (17.27)). The only hope of resolving this difficulty seems to be to resort to methods of numerical integration which take into account the stochastic character of distributions of microproperties in such large systems. The *Monte Carlo methods* are the principal and most well known methods for calculations of this type.

Monte Carlo methods employ random sampling, much in the spirit of the gambling site which inspired their name. They are best appreciated as a method of integration, which is in fact how they are generally employed. Thus, to evaluate an integral of a smooth function $f(x)$,

$$I = \int_a^b f(x) \, dx,$$

we observe that

$$I = (b - a) \langle f \rangle,$$

$\langle f \rangle$ denoting the average of f over interval $[a, b]$. The primitive version of the Monte Carlo method involves evaluating f at a large number K

of randomly generated points $x_i \in [a, b]$ and then

$$\langle f \rangle \approx \frac{1}{K} \sum_{i=1}^{K} f(x_i). \tag{17.46}$$

For smooth f, this approximation converges at a rate $\mathcal{O}(\sqrt{K})$ and, thus, is impractical for the huge systems encountered in statistical mechanics.

One way to accelerate the convergence is to choose a nonnegative function $g(x)$ and write

$$I = \int_a^b \frac{f(x)}{g(x)} g(x)\, dx = \int_a^b \frac{f\big(x(h)\big)}{g\big(x(h)\big)}\, dh,$$

where $h'(x) = g(x)$. Then

$$I \approx \frac{1}{K} \sum_{i=1}^{K} \frac{f\big(x(h_i)\big)}{g\big(x(h_i)\big)}, \tag{17.47}$$

the h_i being random values of $h(x)$ distributed over $[a, b]$. Because we do not, in general, have a priori knowledge on how to choose $g(x)$ or $h(x)$, this biased Monte Carlo method may also be infeasible. We will discuss an alternative approach which overcomes this problem.

17.8.1 The Partition Function for a Canonical Ensemble

Returning to (17.27), we consider a Hamiltonian system of N identical particles of mass m and we rewrite the partition function in the form

$$\Omega = \int_{\mathcal{H}} e^{-H(q,p)/kT}\, dq\, dp = \int_{\mathcal{H}} e^{-(V(q)+p^2/2m)/kT}\, dq\, dp$$

$$= \int_{\mathcal{H}_q} e^{-V(q)/kT}\, dq \int_{\mathcal{H}_p} e^{-p^2/2mkT}\, dp,$$

where the first integral corresponds to the *configurational part* of the system and the second integral corresponds to *the kinetic part*, and we have partitioned $\mathcal{H}$ as $\mathcal{H}_q \times \mathcal{H}_p$. We dispense with the kinetic part by assuming that the motions are small quasi-static fluctuations ($p \approx 0$), so $\exp(-p^2/2mkT) \approx 1$ or we can use the ideal gas assumption which asserts that there are no interactions of particles and that $V(q)$ is zero. In that case,

$$\Omega_{\text{ideal}} = \int_{\mathcal{H}_p} e^{-p^2/2mkT}\, dp = \frac{V^N}{N!\, h^{3N}} (2\pi mkT)^{3N/2}, \qquad (17.48)$$

V being the fixed volume occupied by the N particles. We must then compute

$$\Omega = \Omega_{\text{ideal}} \int_{\mathcal{H}_q} e^{-V(q)/kT}\, dq. \qquad (17.49)$$

The idea behind the use of Monte Carlo methods to evaluate Ω is a straightforward application of (17.46) or (17.47): sample randomly values of the integrand $\exp(-V(q)/kT)$ over $\mathcal{H}_q$ and sum up the approximations. Unfortunately, even with these simplifications, the standard Monte Carlo method is not feasible because of the large number of sample sites that have effectively zero effect on the value of the integral. An alternative is needed.

17.8.2 The Metropolis Method

The Metropolis method provides a version of the Monte Carlo method that provides the means for sampling over the phase space in a way that uses samples which are likely to affect the ensemble average most significantly (see Metropolis et al. [60]). The Metropolis method generates a *Markov chain*, i.e., a sequence of trial samples such that:

- Each trial in the sequence depends only on the preceding trial sample and not on previous samples.

- Each trial belongs to a finite set of possible outcomes.

The goal is to use the Boltzmann factor $\exp(-V(q)/kT)$ as the probability of obtaining a given configuration of the system.

The Metropolis scheme involves generating a chain of configurations $(q_1, q_2, \ldots, q_N)$, calculating $V(q_i)$ for each configuration, and assigning a higher probability to states with lower energy than those with higher energy. For each accepted configuration the value of the desired property (e.g., free energy $A(q_i)$) is calculated and the average over K steps approximates the ensemble average ($\langle A \rangle \approx K^{-1} \sum_{i=1}^{K} A(q_i)$).

Beginning with an arbitrary configuration (a set of values of the coordinate q), a new configuration is generated by random changes (incrementations) of the trial configuration components. The energy of the new configuration is calculated using an atomic or molecular potential function selected for the system as done in molecular dynamics (see Section 16.9). If the new energy is lower than the preceding configuration, it is accepted. If the new energy is higher, the Boltzmann factor of the difference is calculated:

$$ B = \exp\left(-\frac{V(q_{\text{new}}) - V(q_{\text{old}})}{kT} \right). $$

We then choose a random number $r \in [0, 1]$. If $r > B$, the new configuration is rejected. If $r \leq B$, then the move is accepted and the new configuration is accepted as the next state. Thus, the Metropolis method will admit moves to states of higher energy, but the smaller the energy in such a move, the higher the likelihood it will be accepted. The Boltzmann factor B is actually the ratio of the probabilities of successive steps, so this ratio declines in magnitude as a new configuration is accepted. The final step yields an approximation of the probability distribution (the distribution function) necessary for the calculation of an approximation of the ensemble average of various quantities of interest.

17.9 Kinetic Theory: Boltzmann's Equation of Nonequilibrium Statistical Mechanics

17.9.1 Boltzmann's Equation

We shall now depart from the framework of equilibrium statistical mechanics and undertake a brief venture into the so-called kinetic theory of gases and fluids in which the possibility of collisions and rebounding of particles moving at high speeds is admitted. Through a series of simplifying approximations, we can arrive at Boltzmann's equation, which, remarkably, provides one connection of models of statistical mechanics to the macroscopic models of the continuum mechanics of compressible fluids and gases.

We begin by returning to the Liouville equation (16.22) but now restrict ourselves to the density function $\varrho = \varrho_1(\mathbf{q}, \mathbf{p}, t)$ of a single particle, which may collide with another particle of equal mass m. Now ϱ_1 is a function of only seven variables: three components of the position vector $\mathbf{x} = \mathbf{q}$, three velocity components $\mathbf{w} = \mathbf{p}/m$, and time t; and we absorb m in the definition of $\mathbf{F}$. We write $\varrho_1 = \varrho_1(\mathbf{x}, \mathbf{w}, t)$. According to (16.22), with L the Liouville operator, we rewrite (16.22) in the form

$$L\varrho_1 = \frac{\partial \varrho_1}{\partial t} + \mathbf{w} \cdot \nabla_{\mathbf{x}}\varrho_1 + \mathbf{F} \cdot \nabla_{\mathbf{w}}\varrho_1 = \left(\frac{\partial \varrho_1}{\partial t}\right)_{\text{coll.}}, \qquad (17.50)$$

where $(\partial \varrho_1/\partial t)_{\text{coll.}}$ is the *collision term* characterizing the effects of a two-particle collision on ϱ. This equation, as noted earlier, is called Boltzmann's equation. It is central to the kinetic theory of gases. The principal issue is to develop a characterization of the collision term as a function of ϱ_1.

The basic collision conditions follow from a number of simplifying assumptions that employ the laws of classical particle dynamics: Particle 1 with position $\mathbf{x}_1$ and velocity $\mathbf{w}_1$ collides with particle 2 with position $\mathbf{x}_2$ and velocity $\mathbf{w}_2$. We assume that the particles are "hard spheres," so there is a perfectly elastic impact, after which particles have velocity $\mathbf{w}_1'$ and $\mathbf{w}_2'$, respectively. Since momentum and energy are conserved, we have

$$m\mathbf{w}_1 + m\mathbf{w}_2 = m\mathbf{w}_1' + m\mathbf{w}_2',$$
$$\tfrac{1}{2}m|\mathbf{w}_1|^2 + \tfrac{1}{2}m|\mathbf{w}_2|^2 = \tfrac{1}{2}m|\mathbf{w}_1'|^2 + \tfrac{1}{2}m|\mathbf{w}_2'|^2.$$

From these relations, one can show that

$$\mathbf{V}' = \mathbf{V} - 2\mathbf{n}\,(\mathbf{V} \cdot \mathbf{n}),$$

where $\mathbf{V}' = \mathbf{w}_1' - \mathbf{w}_2'$, $\mathbf{V} = \mathbf{w}_1 - \mathbf{w}_2$, and $\mathbf{n}$ is a unit vector along $\mathbf{w}_1 - \mathbf{w}_2$. Thus, $\mathbf{V}_n = (\mathbf{V} \cdot \mathbf{n})\mathbf{n}$ changes sign upon impact and the component $\mathbf{V}_t$ in the plane normal to $\mathbf{n}$ remains unchanged.

To proceed, we need to know the probability ϱ_2 of finding the first particle at $\mathbf{x}_1$ with velocity $\mathbf{w}_1$ and the second at $\mathbf{x}_2$ with velocity $\mathbf{w}_2$ at time t: $\varrho_2 = \varrho_2(\mathbf{x}_1, \mathbf{x}_2, \mathbf{w}_1, \mathbf{w}_2, t)$. If σ is the diameter of the spherical particle, one can argue that (see, e.g., Cercignani [47, pp. 4–6])

$$\left(\frac{\partial \varrho_1}{\partial t}\right)_{\text{coll.}} = (N-1)\sigma^2 \int_{\mathbb{R}^3} \int_{\mathcal{B}} \varrho_2(\mathbf{x}_1, \mathbf{x}_1 + \sigma \mathbf{n}, \mathbf{w}_1, \mathbf{w}_2, t)$$
$$(\mathbf{w}_2 - \mathbf{w}_1) \cdot \mathbf{n} \, dw_2 \, dn,$$

where N is the total number of particles and $\mathcal{B}$ is the sphere corresponding to $\mathbf{V} \cdot \mathbf{n} = 0$. To simplify this relation further and eliminate the need for determining ϱ_2 in addition to ϱ_1, we invoke the *assumption of molecular chaos*, to wit:

$$\varrho_2(\mathbf{x}_1, \mathbf{x}_2, \mathbf{w}_1, \mathbf{w}_2, t) = \varrho_1(\mathbf{x}_1, \mathbf{w}_1, t) \, \varrho_1(\mathbf{x}_2, \mathbf{w}_2, t).$$

The argument in support of this assumption goes like this: Collisions between two randomly selected particles are rare events, so the probability of finding particle 1 at $\mathbf{x}_1$ with velocity $\mathbf{w}_1$ and particle 2 at $\mathbf{x}_2$ with velocity $\mathbf{w}_2$ is the product of the individual probability densities. Collecting these results, we have (cf. Cercignani [47, p. 6])

$$L\varrho_1 = N\sigma^2 \int_{\mathbb{R}^3} \int_{\mathcal{B}^-} \left(\varrho_1(\mathbf{x}_1, \mathbf{w}_1', t) \, \varrho_1(\mathbf{x}_2, \mathbf{w}_2', t) \right.$$
$$\left. - \varrho_1(\mathbf{x}_1, \mathbf{w}_1, t) \, \varrho_1(\mathbf{x}_2, \mathbf{w}_2, t) \right) \mathbf{V} \cdot \mathbf{n} \, dw_2 \, dn,$$
$$(17.51)$$

where, again, $\mathbf{V} = \mathbf{w}_2 - \mathbf{w}_1$, and $\mathcal{B}^-$ is the hemisphere, $\mathbf{V} \cdot \mathbf{n} \leq 0$.

We now look toward generalizations of (17.51) in which ϱ_1 is replaced by multiples of ϱ_1, denoted simply f, which represents the number or mass density of the phase space $\mathcal{H}$ and $N\sigma^2\mathbf{V} \cdot \mathbf{n}$ is replaced by a kernel $\mathcal{B}(\theta, \mathbf{V})$ depending on $\mathbf{V}$ and the solid angle θ. Then (17.51) reduces to

$$\boxed{Lf = Q(f, f),} \qquad (17.52)$$

where $Q(f, f)$ is the collision integral,

$$Q(f, f) = \int_{\mathbb{R}^3} \int_{B^-} (f'f'_* - ff_*) \, B(\theta, \mathbf{V}) \, dw_* \, d\theta \, d\alpha. \qquad (17.53)$$

Here $f' = f(\mathbf{x}, \mathbf{w}', t)$, $f'_* = f(\mathbf{x}, \mathbf{w}'_*, t)$, and α is the angle that, given θ, defines n.

Boltzmann's equation (17.52) is clearly highly nonlinear in the density function f. Its numerical solution is the subject of a growing literature on gas dynamics and statistical mechanics.

17.9.2 Collision Invariants

The particular density of functions f that make the collision integral Q vanish are of special interest. First, we note that any function $\varphi(\mathbf{w})$ such that

$$\int_{\mathbb{R}^3} \varphi(\mathbf{w}) \, Q(f, g) \, dw = 0, \qquad (17.54)$$

for all densities f, g is called a *collision invariant*, as it represents the rate of change of the average value of φ due to collisions. It can be shown that a continuous function φ has property (17.54) if and only if

$$\varphi(\mathbf{w}) = \alpha + \boldsymbol{\beta} \cdot \mathbf{w} + c\mathbf{w} \cdot \mathbf{w}, \qquad (17.55)$$

where α, $\boldsymbol{\beta}$, and c are constants ($c > 0$). The five functions $\Gamma_1 = 1$, $(\Gamma_2, \Gamma_3, \Gamma_4) = (\mathbf{w}_1, \mathbf{w}_2, \mathbf{w}_3)$, and $\Gamma_5 = |\mathbf{w}|^2$ are called the *collision invariants*. When $\varphi = \ln f$, the Boltzmann inequality holds:

$$\int_{\mathbb{R}^3} \ln f \, Q(f, f) \, dw \leq 0, \qquad (17.56)$$

and the equality holds if and only if $(\ln f)$ is of the form (17.55), i.e.,

$$f = \exp(\alpha + \boldsymbol{\beta} \cdot \mathbf{w} + |\mathbf{w}|^2),$$

which is equivalent to

$$f = A \exp\left(-\beta |\mathbf{w} - \mathbf{v}|^2\right), \tag{17.57}$$

A, β, and $\mathbf{v}$ being a new set of constants determined by α, β, and c. The function f in (17.57) is called the *Maxwell distribution* or the *Maxwellian*. Returning to the original problem of determining the particular density functions f that satisfy $Q(f, f) = 0$, one can now easily show that all such functions are Maxwellians. The arguments surrounding inequality (17.56) underscore the *Boltzmann H-theorem* whereby the integral appearing in (17.56) is the time derivative of a function $\mathcal{H} = \int_{\mathbb{R}^3} f \ln f \, dw$. The Boltzmann inequality (17.56) implies that $\mathcal{H}$ is a decreasing quantity unless f is Maxwellian, in which case $d\mathcal{H}/dt = 0$. This, for space-homogeneous systems, characterizes irreversibility; see Cercignani [47, p. 23] or McQuarrie [59, pp. 413-415].

17.9.3 The Continuum Mechanics of Compressible Fluids and Gases: The Macroscopic Balance Laws

We now make the full circle: the derivation of the balance laws of the macrosystem from the equations of kinetic theory, that is, from Boltzmann's equation (17.52). Now let f be the (expected) mass density in phase space which is a solution to Boltzmann's equation,

$$\frac{\partial f}{\partial t} + \mathbf{w} \cdot \nabla_{\mathbf{x}} f + \mathbf{F} \cdot \nabla_{\mathbf{w}} f = Q(f, f). \tag{17.58}$$

Let Γ_α denote the collision invariants described earlier ($\Gamma_1 = 1$, $(\Gamma_2, \Gamma_3, \Gamma_4) = \mathbf{w}$, $\Gamma_5 = |\mathbf{w}|^2$). Then

$$\int_{\mathbb{R}^3} \Gamma_\alpha(\mathbf{w}) \, Q(f, f) \, dw = 0, \qquad \alpha = 1, \ldots, 5. \tag{17.59}$$

Introducing (17.58) into (17.59) and changing the order of integration gives

$$\frac{\partial}{\partial t} \int_{\mathbb{R}^3} \Gamma_\alpha f \, dw + \nabla_{\mathbf{x}} \cdot \int_{\mathbb{R}^3} \mathbf{w} \, \Gamma_\alpha f \, dw + \int_{\mathbb{R}^3} \mathbf{F} \cdot \nabla_{\mathbf{w}} \Gamma_\alpha f \, dw = 0,$$
$$\alpha = 1, \ldots, 5. \tag{17.60}$$

We now assign the Γ_α their respective values (functions) as collision invariants. To interpret the resulting conditions, we introduce the following macroscopic variables:

$$\varrho(\mathbf{x}, t) = \int_{\mathbb{R}^3} f(\mathbf{x}, \mathbf{w}, t)\, dw \qquad\qquad = \text{mass density,}$$

$$\mathbf{v}(\mathbf{x}, t) = \frac{1}{\varrho} \int_{\mathbb{R}^3} \mathbf{w}\, f(\mathbf{x}, \mathbf{w}, t)\, dw \qquad = \text{bulk velocity,}$$

$$E(\mathbf{x}, t) = \frac{1}{2} \int_{\mathbb{R}^3} |\mathbf{w}|^2 f(\mathbf{x}, \mathbf{w}, t)\, dw = \text{total energy per unit volume,}$$

$$\mathbf{m}(\mathbf{x}, t) = \int_{\mathbb{R}^3} \mathbf{w} \otimes \mathbf{w}\, f(\mathbf{x}, \mathbf{w}, t)\, dw = \text{momentum flux,}$$

$$\mathbf{r}(\mathbf{x}, t) = \int_{\mathbb{R}^3} \mathbf{w}\, |\mathbf{w}|^2 f(\mathbf{x}, \mathbf{w}, t)\, dw = \text{energy flux.}$$

$$(17.61)$$

In addition, we define the phase velocity $\mathbf{u}$ as the perturbation of the macrovelocity field $\mathbf{w}$ relative to the bulk velocity $\mathbf{v}$:

$$\mathbf{u} = \mathbf{w} - \mathbf{v},$$

so that $(17.61)_4$ can be written

$$\mathbf{m} = \varrho\, \mathbf{v} \otimes \mathbf{v} + \mathbf{T}, \qquad\qquad (17.62)$$

where

$$\mathbf{T} = \int_{\mathbb{R}^3} \mathbf{u} \otimes \mathbf{u}\, f\, dw. \qquad\qquad (17.63)$$

The tensor $\mathbf{T}$ is the (Cauchy) stress tensor, representing the momentum flow at any interior surface of the medium. Similarly, the total energy can be written as

$$E = \int_{\mathbb{R}^3} \frac{1}{2} |\mathbf{u} + \mathbf{v}|^2 f\, dw,$$

or, since $\int_{\mathbb{R}^3} \mathbf{u} f \, dw = 0$,

$$E = \frac{1}{2}\varrho|\mathbf{v}|^2 + \varrho e, \tag{17.64}$$

where e is the internal energy per unit mass:

$$\varrho e = \frac{1}{2}\int_{\mathbb{R}^3} |\mathbf{u}|^2 f \, dw. \tag{17.65}$$

Finally, we have

$$\mathbf{r} = \int_{\mathbb{R}^3} |\mathbf{u} + \mathbf{v}|^2 (\mathbf{u} + \mathbf{v}) f \, dw$$

$$= \varrho \mathbf{v}\left(\frac{1}{2}|\mathbf{v}|^2 + \varrho e\right) + \int_{\mathbb{R}^3} (\mathbf{u} \otimes \mathbf{u})\,\mathbf{v} f \, dw + \mathbf{q},$$

or

$$\mathbf{r} = \varrho \mathbf{v} E + \mathbf{T}\mathbf{v} + \mathbf{q}, \tag{17.66}$$

where $\mathbf{q}$ is the heat flux vector

$$\mathbf{q} = \frac{1}{2}\int_{\mathbb{R}^3} \mathbf{u}\,|\mathbf{u}|^2 f \, dw. \tag{17.67}$$

Returning now to (17.60), we first set $\alpha = 1$, $\Gamma_1 = 1$ and obtain the equation of mass conservation ($\nabla_{\mathbf{x}} \equiv \nabla$),

$$\frac{\partial \varrho}{\partial t} + \nabla \cdot (\varrho \mathbf{v}) = 0, \tag{17.68}$$

Next setting $(\Gamma_2, \Gamma_3, \Gamma_4) = \mathbf{w}$ and denoting $\int_{\mathbb{R}^3} \mathbf{F} \cdot \nabla_{\mathbf{w}} \mathbf{w} f \, dx = \varrho \mathbf{f}$, we obtain

$$\frac{\partial}{\partial t}(\varrho \mathbf{v}) + \nabla \cdot (\varrho \mathbf{v} \otimes \mathbf{v} + \mathbf{T}) = \varrho \mathbf{f}. \tag{17.69}$$

Finally, with $\Gamma_5 = |\mathbf{w}|^2$ and $\int_{\mathbb{R}^3} \mathbf{F} \cdot \nabla_{\mathbf{w}} |\mathbf{w}|^2 f\ dx = \varrho \mathbf{f} \cdot \mathbf{v}$, we get

$$\frac{\partial}{\partial t} \left(\tfrac{1}{2} \varrho |\mathbf{v}|^2 + \varrho e \right) + \nabla \cdot \left(\varrho \mathbf{v} (\tfrac{1}{2} |\mathbf{v}|^2 + e) + \mathbf{T}\mathbf{v} + \mathbf{q} \right) = \varrho \mathbf{f} \cdot \mathbf{v}.$$

$$(17.70)$$

We recognize (17.69) as the macroscopic equation of balance of linear momentum for a continuous media and (17.70) as the statement of the principle of conservation of energy for the continuum. In these equations, $\mathbf{f}$ is assumed to be given as a body force per unit mass. If the medium $\mathcal{B}$ occupies a current configuration $\Omega_t \subset \mathbb{R}^3$ at time t, (17.68) and (17.69) assert that the total mass $\mathcal{M}(\mathcal{B})$ of the body is conserved and the time rate of change of the linear momentum $\mathcal{I}(\mathcal{B})$ is balanced by the total force acting on the body, where

$$\mathcal{M}(\mathcal{B}) = \int_{\Omega_t} \varrho\ dx \quad \text{and} \quad \mathcal{I}(\mathcal{B}) = \int_{\Omega_t} \varrho \mathbf{v}\ dx.$$

Likewise, (17.70) establishes that energy,

$$E(\mathcal{B}) = \int_{\Omega_t} \left(\tfrac{1}{2} \varrho |\mathbf{v}|^2 + \varrho e \right) dx,$$

is conserved, in the sense that its time rate of change is balanced by the power plus the rate of heating.

Exercises

EXERCISES FOR PART I

Things You Should Know to Read This Book An introductory understanding of:

- Linear algebra and matrix theory

- Vector calculus

- Index notation

- Introductory real analysis

Brief Review of Index Notation and Symbolic Notation

- Let e_i, $i = 1, 2, 3$, be an orthonormal basis, i.e.,

$$e_i \cdot e_j = \delta_{ij} = \begin{cases} 1, & \text{if } i = j, \\ 0, & \text{if } i \neq j. \end{cases}$$

- Let a be a vector: $a = a_i\, e_i$ (repeated indices are summed). Two vectors a and b are equal (i.e., $a = b$) if $a_i = b_i$, $i = 1, 2, 3$.

- Cross product: $a \times b = \varepsilon_{ijk} a_i b_j e_k$

$$\varepsilon_{ijk} = \begin{cases} 1, & \text{if } ijk = \text{even permutation}, \\ -1, & \text{if } ijk = \text{odd permutation}, \\ 0, & \text{if } ijk \text{ is not a permutation}. \end{cases}$$

- Nabla: $\nabla = e_k \dfrac{\partial}{\partial x_k}$.

- Divergence of a vector (denoted div v or $\nabla \cdot v$):

$$\nabla \cdot v = e_k \frac{\partial}{\partial x_k} \cdot v_j e_j = \frac{\partial v_j}{\partial x_k} e_k \cdot e_j = \frac{\partial v_j}{\partial x_k} \delta_{kj} = \frac{\partial v_k}{\partial x_k} (= v_{k,k} \text{ or } \partial_k v_k).$$

- Curl of a vector (denoted curl v or $\nabla \times v$):

$$\nabla \times v = \varepsilon_{ijk} \frac{\partial}{\partial x_i} v_j e_k = \varepsilon_{ijk} v_{j,i} e_k = \varepsilon_{ijk} (\partial_i v_j) e_k.$$

- The following relations hold:

$$\delta_{ii} = 3,$$
$$\delta_{ij} \delta_{jk} = \delta_{ik},$$
$$\varepsilon_{ijk} \varepsilon_{ijm} = 2\delta_{km},$$
$$\varepsilon_{ijk} \varepsilon_{imn} = \delta_{jm} \delta_{kn} - \delta_{jn} \delta_{km},$$
$$\varepsilon_{ijk} \varepsilon_{ijk} = 6.$$

- Typical identities:
  1. $(v \cdot \nabla) v = \frac{1}{2} \nabla (v \cdot v) - v \times (\nabla \times v)$.
  2. $\nabla \times (v \times w) = (w \cdot \nabla) v + v \cdot (\nabla \cdot w) - w \cdot (\nabla \cdot v) - (v \cdot \nabla) w$.

- Tensors:

$e_i \otimes e_j = $ tensor product of e_i and e_j.

$$\mathbf{A} = \text{second-order tensor} = \left(\sum_{i,j} \right) A_{ij} e_i \otimes e_j = A_{ij} \, e_i \otimes e_j,$$

$$(A_{ij} = e_i \cdot \mathbf{A} e_j).$$

$\mathbf{B} = $ third-order tensor $= B_{ijk} \, e_i \otimes e_j \otimes e_k$.

Exercise Set I.1

Kronecker delta and Permutation Symbol

1. Prove the Kronecker delta and permutation symbol identities:

$$\delta_{ii} = 3,$$
$$\delta_{ij}\delta_{jk} = \delta_{ik},$$
$$\varepsilon_{ijk}\varepsilon_{ijm} = 2\delta_{km},$$
$$\varepsilon_{ijk}\varepsilon_{imn} = \delta_{jm}\delta_{kn} - \delta_{jn}\delta_{km},$$
$$\varepsilon_{ijk}\varepsilon_{ijk} = 6.$$

Vectors and Index Notation (using index notation)

2. Prove that

$$(\boldsymbol{v}\cdot\nabla)\boldsymbol{v} = \frac{1}{2}\nabla(\boldsymbol{v}\cdot\boldsymbol{v}) - \boldsymbol{v}\times(\nabla\times\boldsymbol{v}).$$

3. Prove that

$$\nabla\times(\boldsymbol{v}\times\boldsymbol{w}) = (\boldsymbol{w}\cdot\nabla)\boldsymbol{v} + \boldsymbol{v}(\nabla\cdot\boldsymbol{w}) - \boldsymbol{w}(\nabla\cdot\boldsymbol{v}) - (\boldsymbol{v}\cdot\nabla)\boldsymbol{w}.$$

Vectors and Inner Product Spaces (one can consult standard texts on these subjects, such as Oden and Demkowicz [6])

4. Give a complete definition and a nontrivial example of

   (a) a real vector space;

   (b) an inner product space;

   (c) a linear transformation from a vector space U into a vector space V.

Tensors

5. Let V be an inner product space. A tensor is a linear transformation from V into V. If $\mathbf{T}$ is a tensor, $\mathbf{T}v$ denotes the image of the vector v in V:

$$\mathbf{T}v = \mathbf{T}(v) \in V.$$

Show that the class $L(V, V)$ of all linear transformations of V into itself is also a vector space with vector addition and scalar multiplication defined as follows:

$$\mathbf{S}, \mathbf{T} \in L(V, V):$$
$$\mathbf{S} + \mathbf{T} = \mathbf{R} \Leftrightarrow \mathbf{R}v = \mathbf{S}v + \mathbf{T}v,$$
$$\alpha\mathbf{S} = \mathbf{R} \Leftrightarrow \mathbf{R}v = \alpha(\mathbf{S}v),$$
$$\forall v \in V, \forall \alpha \in \mathbb{R}$$
$$(\mathbf{0} \in L(V, V): \mathbf{0}v = \mathbf{0} \in V).$$

Tensor Product

6. The *tensor product* of two vectors a and b is the tensor, denoted $a \otimes b$, that assigns to each vector c the vector $(b \cdot c)a$; that is,

$$(a \otimes b)c = (b \cdot c)a.$$

6.1 Show that $a \otimes b$ is a tensor and that

$$(a \otimes b)^T = b \otimes a.$$
$$(a \otimes b)(c \otimes d) = (b \cdot c)(a \otimes d).$$

6.2 If $\{e_1, e_2, e_3\}$ is an orthonormal basis ($e_i \cdot e_j = \delta_{ij}$, $1 \le i, j \le 3$), $\|e_i\|^2 = e_i \cdot e_i = 1$), then show that

$$(e_i \otimes e_i)(e_j \otimes e_j) = \begin{cases} \mathbf{0}, & \text{if } i \ne j \\ e_i \otimes e, & \text{if } i = j \end{cases} = \text{``}\delta_{ij} e_i \otimes e_j\text{''}.$$

6.3 For an arbitrary tensor **A**, and for the orthonormal basis $\{e_1, e_2, e_3\}$,

$$\mathbf{A} = \sum_{i,j} A_{ij}\, e_i \otimes e_j,$$

where

$$A_{ij} = e_i \cdot \mathbf{A}e_j.$$

The array $[A_{ij}]$ is the matrix characterizing **A** for this particular choice of a basis for $V (= \mathbb{R}^3)$. If **A** and **B** are two ("second-order") tensors and $[A_{ij}]$, $[B_{ij}]$ are their matrices corresponding to a basis $\{e_1, e_2, e_3\}$ of V, define (construct) the rules of matrix algebra:

(a) $\mathbf{A} + \mathbf{B} = \mathbf{C} \Rightarrow [A_{ij}] + [B_{ij}] = [?]$;

(b) $\mathbf{AB} = \mathbf{C}\ (\mathbf{AB} = \mathbf{A} \circ \mathbf{B})$;

(c) $\mathbf{A0} = \mathbf{C}\ (\mathbf{C} =?)\ (\mathbf{0} = \text{the } zero\ element \text{ of } L(V, V))$;

(d) $\mathbf{AC} = \mathbf{I}\ (\mathbf{I}$ is the $identity\ tensor$: $\mathbf{AI} = \mathbf{IA} = \mathbf{A}$, and $\mathbf{C} = \mathbf{A}^{-1})$;

(e) $\mathbf{A}^T = \mathbf{C}\ (\mathbf{C} =?)\ (\mathbf{A}^T$ is the $transpose$ of **A**: it is the unique tensor such that $\mathbf{A}v \cdot u = v \cdot \mathbf{A}^T u$, "·" being the vector inner product in $\mathbb{R}^3$).

7. The real inner product ("dot product") of two vectors $u, v \in V = \mathbb{R}^3$ is denoted $u \cdot v$. It is a symmetric, positive-definite, bilinear form on V. If $u = \sum_{i=1}^{3} u_i e_i$ and $v = \sum_{i=1}^{3} v_i e_i$, for an orthonormal basis $\{e_i\}$, then $u \cdot v = \sum_{i=1}^{3} u_i v_i$. Moreover, the (Euclidean) norm of u (v) is $\|u\| = \sqrt{u \cdot u}$ $(\|v\| = \sqrt{v \cdot v})$.

The space $L(V, V)$ of second-order tensors can be naturally equipped with an inner product as well, and hence a norm. The construction is as follows:

i) The $trace$ of the $tensor\ product$ of two vectors u and v is the linear operation

$$\text{tr}(u \otimes v) \overset{def}{=} u \cdot v.$$

Likewise, the *trace* of a tensor $\mathbf{A} \in L(V,V)$ is defined by

$$\text{tr}\mathbf{A} = \text{tr}\left(\sum_{ij} A_{ij}\, \mathbf{e}_i \otimes \mathbf{e}_j\right) = \sum_{ij} A_{ij}\, \text{tr}\, \mathbf{e}_i \otimes \mathbf{e}_j$$
$$= \sum_{ij} A_{ij}\delta_{ij} = \sum_{i} A_{ii}.$$

ii) The trace of the composition of two tensors $\mathbf{A}, \mathbf{B} \in L(V,V)$ is then
$$\text{tr}(\mathbf{A}\mathbf{B}) = \sum_{ij} A_{ij}B_{ji}.$$

We denote

$$\mathbf{A} : \mathbf{B} = \text{tr}(\mathbf{A}^T\mathbf{B}) = \sum_{ij} A_{ij}B_{ij}.$$

(a) Show that $\mathbf{A} : \mathbf{B}$ (the operation ":") defines an inner product on $L(V,V)$.

(b) Define the associated norm $\|\mathbf{A}\|$ of $\mathbf{A} \in L(V,V)$.

(c) Show that
$$\text{tr}\,\mathbf{A} = \text{tr}\,\mathbf{A}^T.$$

(d) Show that (trivially)

$$\mathbf{I} : \mathbf{A} = \text{tr}\,\mathbf{A} \qquad (\mathbf{I} = \text{identity tensor}),$$
$$(\mathbf{u} \otimes \mathbf{v}) : (\mathbf{q} \otimes \mathbf{p}) = (\mathbf{u} \cdot \mathbf{q})(\mathbf{v} \cdot \mathbf{p}).$$

8. Let φ, $\mathbf{u}$, $\mathbf{v}$, and $\mathbf{A}$ be smooth fields; φ scalar, $\mathbf{u}$ and $\mathbf{v}$ vectors, and $\mathbf{A}$ tensor. Show that:

$$\nabla(\varphi\mathbf{v}) = \varphi\nabla\mathbf{v} + \mathbf{v} \otimes \nabla\varphi,$$
$$\text{div}\,(\mathbf{u} \otimes \mathbf{v}) = \mathbf{u}\,\text{div}\,\mathbf{v} + (\text{grad}\,\mathbf{u})\mathbf{v},$$
$$\mathbf{A}\nabla\varphi = \text{div}\,(\varphi\mathbf{A}) - \varphi\,\text{div}\,\mathbf{A}.$$

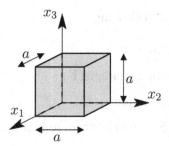

Figure E.1: The cube $\Omega_0 = (0, a)^3$.

Exercise Set I.2

Kinematics of Continuous Media

1. The reference configuration of a deformable body $\mathcal{B}$ is the cube $\Omega_0 = (0, a)^3$ with the origin of the spatial and material coordinates at the corner, as shown in Fig E.1. Consider a motion φ of the body defined by

$$\varphi(\boldsymbol{X}) = \sum_{i=1}^{3} \varphi_i(\boldsymbol{X})e_i,$$

where

$$\varphi_1(\boldsymbol{X}) = X_1 + \frac{\Delta}{2}\left(\frac{X_2}{a}\right)^2,$$
$$\varphi_2(\boldsymbol{X}) = X_2,$$
$$\varphi_3(\boldsymbol{X}) = X_3.$$

where Δ is a real number (a parameter possibly depending on time t).

For this motion,

(a) Sketch the deformed shape (i.e., sketch the current configuration) in the X_1–X_2 plane for $\Delta = a$.

Then compute the following:

(b) the displacement field u,

(c) the deformation gradient $\mathbf{F}$,

(d) the deformation tensor $\mathbf{C}$,

(e) the Green–St. Venant strain tensor $\mathbf{E}$,

(f) the extensions e_i, $i = 1, 2, 3$,

(g) $\sin \gamma_{12}$, where γ_{12} is the shear in the X_1–X_2 plane.

Determinants

2. Let $\mathbf{A} \in L(V, V)$ be a second order tensor with a matrix $[A_{ij}]$ relative to a basis $\{e_i\}_{i=1}^3$ (i.e., the A_{ij} are components of $\mathbf{A}$ with respect to $\{e_i \otimes e_j\}$). For $n = 3$, the determinant of $[A_{ij}]$ is defined by

$$\det[A_{ij}] = \frac{1}{6} \sum_{\substack{ijk \\ rst}} \varepsilon_{ijk} \varepsilon_{rst} A_{ir} A_{js} A_{kt},$$

where

$$\varepsilon_{ijk} = \begin{cases} 1 & \text{if } \{i, j, k\} \text{ is an even permutation of } \{i, j, k\}, \\ -1 & \text{if } \{i, j, k\} \text{ is an odd permutation of } \{i, j, k\}, \\ 0 & \text{if } \{i, j, k\} \text{ is not a permutation of } \{i, j, k\} \end{cases}$$

$$\text{(i.e., if at least two indices are equal).}$$

The determinant of the tensor $\mathbf{A}$ is defined as the determinant of its matrix components A_{ij}:

$$\det \mathbf{A} = \det[A_{ij}].$$

This definition is *independent of the choice of basis* $\{e_i\}$ (i.e., $\det \mathbf{A}$ is a property of $\mathbf{A}$ *invariant* under changes of basis).

Show that

(a) $\det(\mathbf{AB}) = (\det \mathbf{A})(\det \mathbf{B})$ (for $n=3$ is sufficient).
Hint:

$$\det \mathbf{A} = \frac{1}{6} \varepsilon_{k\ell m} \varepsilon_{abc} A_{ka} A_{\ell b} A_{mc},$$

$$\det \mathbf{B} = \frac{1}{6} \varepsilon_{k\ell m} \varepsilon_{def} B_{kd} B_{\ell e} B_{mf}.$$

But also

$$\varepsilon_{k\ell m} \det \mathbf{A} = \varepsilon_{abc} A_{ka} A_{\ell b} A_{mc},$$
$$\varepsilon_{k\ell m} \det \mathbf{B} = \varepsilon_{def} B_{kd} B_{\ell e} B_{mf}$$
(for example: $\det \mathbf{A} = \varepsilon_{abc} A_{1a} A_{2b} A_{3c}$).

Note also that $\varepsilon_{k\ell m} \varepsilon_{k\ell m} = 6$.

(b) $\det \mathbf{A}^T = \det \mathbf{A}$.

Cofactor and Inverse

3. For $\mathbf{A} \in L(V, V)$ and A_{ij} the components of $\mathbf{A}$ relative to a basis $\{\mathbf{e}_i\}$, $i = 1, 2, \ldots, n$, let A_{ij}^1 be the elements of a matrix of order $(n - 1)$ obtained by deleting the ith row and jth column of $[A_{ij}]$. The scalar
$$d_{ij} = (-1)^{i+j} \det[A_{ij}^1]$$
is called the (i, j)-*cofactor* of $[A_{ij}]$ and the matrix of cofactors = Cof $\mathbf{A} = [d_{ij}]$ is called the *cofactor matrix* of $\mathbf{A}$.

(a) Show that
$$\mathbf{A}(\text{Cof } \mathbf{A})^T = (\det \mathbf{A})\mathbf{I}$$
(it is sufficient to show this for $n = 3$ by construction).

(b) Show (trivially) that for $\mathbf{A}$ invertible,
$$\mathbf{A}^{-1} = (\det \mathbf{A})^{-1}(\text{Cof } \mathbf{A})^T.$$

(c) For $n = 3$, show that
$$(\text{Cof } \mathbf{A})_{kt} = \frac{1}{2}\varepsilon_{ijk}\varepsilon_{rst}A_{ir}A_{js}, \qquad 1 \leq k, t \leq 3.$$

4. Derive (1.25) using the facts that

$$\det \mathbf{F} = \frac{1}{6}\varepsilon_{ijk}\varepsilon_{rst}F_{ir}F_{js}F_{kt},$$

$$(\operatorname{Cof}\mathbf{F})_{ir} = \frac{1}{2}\varepsilon_{ijk}\varepsilon_{rst}F_{js}F_{kt}.$$

Proceed as follows:

(a) First, show that

$$\overline{\det \mathbf{F}} = \operatorname{Cof}\mathbf{F} : \dot{\mathbf{F}}.$$

(b) Next, show that $\dot{\mathbf{F}} = \mathbf{L}\mathbf{F}$ and, therefore, that $\overline{\det \mathbf{F}} = (\operatorname{Cof}\mathbf{F})^{T}\mathbf{F} : \mathbf{L} = (\det\mathbf{F})\mathbf{I} : \mathbf{L}$, which implies (1.25).

Orthogonal Transformation

5. A tensor $\mathbf{Q} \in L(V,V)$ is orthogonal if it preserves inner products, in the sense that
$$\mathbf{Qu} \cdot \mathbf{Qv} = \boldsymbol{u} \cdot \boldsymbol{v}.$$
$\forall \boldsymbol{u}, \boldsymbol{v} \in V(\approx \mathbb{R}^3)$. Show that a necessary and sufficient condition that $\mathbf{Q}$ be orthogonal is that

$$\mathbf{Q}\mathbf{Q}^T = \mathbf{Q}^T\mathbf{Q} = \mathbf{I},$$

or, equivalently,

$$\mathbf{Q}^T = \mathbf{Q}^{-1}.$$

6. Confirm that $\det\mathbf{A}$ is an invariant of $\mathbf{A}$ in the following sense: If

$$\mathbf{A} = \sum_{i,j} A_{ij}\,\boldsymbol{e}_i \otimes \boldsymbol{e}_j = \sum_{i,j}\bar{A}_{ij}\,\bar{\boldsymbol{e}}_i \otimes \bar{\boldsymbol{e}}_j,$$

where

$$\bar{\boldsymbol{e}}_i = \mathbf{Q}\boldsymbol{e}_i, \ i = 1,2,3, \ \mathbf{Q} = \text{an orthogonal tensor},$$

then

$$\det[A_{ij}] = \det[\bar{A}_{ij}].$$

Review of Vector and Tensor Calculus

7. Let U be a normed space and let $f : U \to \mathbb{R}$ be a function from U to $\mathbb{R}$ (such a function is called a *functional* on U). The space $U' = L(U, \mathbb{R})$ is called the *dual* of U. We say that f is differentiable at $\boldsymbol{u} \in U$ if there exists a linear function $\boldsymbol{g} \in U'$ such that

$$\lim_{\theta \to 0} \frac{1}{\theta} \left(\mathbf{f}(\boldsymbol{u} + \theta \boldsymbol{v}) - f(\boldsymbol{u}) \right) = \langle \boldsymbol{g}, \boldsymbol{v} \rangle,$$

$\forall v \in U$. Here $\langle \cdot, \cdot \rangle$ denotes "duality pairing", meaning the number $\langle \boldsymbol{g}, \boldsymbol{v} \rangle \in \mathbb{R}$ depends linearly on $\boldsymbol{g}$ and on $\boldsymbol{v}$. In particular, $g(\boldsymbol{v}) \equiv \langle \boldsymbol{g}, \boldsymbol{v} \rangle$. We write

$$\boldsymbol{g} = Df(\boldsymbol{u}),$$

and call $Df(\boldsymbol{u})$ the derivative (or the Gâteaux or variational derivative) of f at $\boldsymbol{u}$.

Let φ be a scalar-valued function defined on the set $\mathcal{S}$ of invertible tensors $\mathbf{A} \in L(U, U)$ (i.e. $\varphi : \mathcal{S} \to \mathbb{R}$) defined by

$$\varphi(\mathbf{A}) = \det \mathbf{A}.$$

Show that (it is sufficient to consider the 3D case), $\forall \mathbf{V} \in L(U, U)$,

$$D\varphi(\mathbf{A}) : \mathbf{V} = (\det \mathbf{A}) \mathbf{V}^T : \mathbf{A}^{-1}.$$

Hint: Note that

$$\det(\mathbf{A} + \theta \mathbf{V}) = \det((\mathbf{I} + \theta \mathbf{V} \mathbf{A}^{-1}) \mathbf{A})$$
$$= \det \mathbf{A} \det(\mathbf{I} + \theta \mathbf{V} \mathbf{A}^{-1}),$$

and that

$$\det(\mathbf{I} + \theta \mathbf{B}) = 1 + \theta \operatorname{tr} \mathbf{B} + O(\theta^2).$$

8. Let Ω be an open set in $\mathbb{R}^3$ and φ be a smooth function mapping Ω into $\mathbb{R}$. The vector $\boldsymbol{g}$ with the property

$$D\varphi(\boldsymbol{x}) \cdot \boldsymbol{v} = \boldsymbol{g}(\boldsymbol{x}) \cdot \boldsymbol{v} \qquad \forall v \in \mathbb{R}^3,$$

is the gradient of v at point $x \in \Omega$. We use the classical notation,

$$g(x) = \nabla\varphi(x).$$

Show that

$$\nabla(\varphi v) = \varphi \nabla v + v \otimes \nabla\varphi,$$
$$\mathrm{div}(\varphi v) = \varphi \, \mathrm{div} \, v + v \cdot \nabla\varphi,$$
$$\mathrm{div}(v \otimes w) = v \, \mathrm{div} \, w + (\nabla v)w.$$

9. Let φ and u be C^2 scalar and vector fields. Show that

   (a) $\mathrm{curl} \, \nabla\varphi = 0$;

   (b) $\mathrm{div} \, \mathrm{curl} \, v = 0$.

10. Let Ω be an open, connected, smooth domain in $\mathbb{R}^3$ with boundary $\partial\Omega$. Let $\mathbf{n}$ be a unit exterior normal to $\partial\Omega$. Recall Green's (divergence) theorem,[1]

$$\int_\Omega \mathrm{div} \, v \, dx = \int_{\partial\Omega} v \cdot \mathbf{n} \, dA,$$

for a vector field v. Show that

$$\int_\Omega \mathrm{div} \, \mathbf{A} \, dx = \int_{\partial\Omega} \mathbf{A}\mathbf{n} \, dA.$$

Hint: Note that for arbitrary vector u,

$$u \cdot \int_{\partial\Omega} \mathbf{A}\mathbf{n} \, dA = \int_{\partial\Omega} (\mathbf{A}^T u) \cdot \mathbf{n} \, dA.$$

11. If (x, y, z) is a Cartesian coordinate system with origin at the corner of a cube $\Omega_0 = (0, 1)^3$ and axes along edges of the cube, compute

$$\int_{\partial\Omega_0} v \cdot \mathbf{n} \, dA_0,$$

where $\partial\Omega_0$ is the exterior surface of Ω_0, $\mathbf{n}$ is the unit exterior normal vector, and v is the field, $v = xe_1 + ye_2 + 3ze_3$.

---

[1] div v = divergence of v = tr $\nabla v = \sum_i \partial v_i / \partial x_i$ if $v = \sum_i v_i e_i$. div $\mathbf{A}$ is the unique vector field such that div $\mathbf{A} \cdot a = \mathrm{div}(\mathbf{A}^T a)$ for all vectors a.

Exercise Set I.3

1. The cube $(0, a)^3$ is the reference configuration Ω_0 of a body sub-jected to "simple shear"

$$u_1 = \frac{\Delta}{a} X_2, \quad u_2 = u_3 = 0, \quad \Delta = \text{constant.}$$

The unit exterior normal to $\partial \Omega_0$ is $\mathbf{n}_0$. The vector $\mathbf{n}_0$ at boundary point $(a, a/2, a/2)$ is mapped into the unit exterior normal $\mathbf{n}$ on $\partial \Omega_t$. Compute $\mathbf{n}$ and sketch the deformed body.

2. Suppose Δ in Problem 1 above is a function of time t. Compute

   (a) the Lagrangian description of the velocity and acceleration fields,

   (b) the Eulerian descriptions of these fields,

   (c) $\mathbf{L} = \text{grad } \boldsymbol{v}$.

3. The following displacement field is considered for a material body:

$$u_1 = 0, \quad u_2 = 2X_1^2, \quad u_3 = 4X_2^2.$$

   (a) Show that this deformation is possible in a continuously deformable body.

   (b) Two material lines, AB and AC are drawn through a point A with coordinates $(1, 1, 1)$ in the reference configuration, so that $\overline{AB} = 0.5\mathbf{e}_2$ and $\overline{AC} = 2.0\mathbf{e}_3$ (i.e. $\overline{AB}$ is of length 0.5 and is initially oriented along the X_2 axis, etc.). Determine the vectors in the current configuration into which these are deformed.

   (c) Determine the change in angle between $\overline{AB}$ and $\overline{AC}$ due to the deformation.

   (d) Determine the extension e_2 at $(1, 1, 1)$.

4. (a) For the deformation described in 3. above, find the principal strains and their directions at the point $(1, 1, 0)$.

(b) Given the material plane P defined by the vectors $2(1,0,0)$ and $3(0,1,1)$ in the reference configuration, find the unit normal to this surface in the current configuration.

5. Suppose that the reference configuration of a body is the right circular cylinder,

$$\overline{\Omega_0} = \Big\{ (r,\theta,z) = \big(r = (X_1^2 + X_2^2)^{1/2},\ \theta = \arctan(X_2/X_1),$$

$$z = X_3\big) : 0 \le r \le r_0,\ 0 \le \theta \le 2\pi,\ 0 \le z \le L \Big\}.$$

The cylinder is subjected to the deformation (motion),

$$x_1 = X_1 \cos(\phi X_3) - X_2 \sin(\phi X_3),$$
$$x_2 = X_1 \sin(\phi X_3) + X_2 \cos(\phi X_3),$$
$$x_3 = X_3,$$

where ϕ is the *angle of twist* per unit length. This describes a *torsional deformation* of the body (since $x_1^2 + x_2^2 = X_1^2 + X_2^2$).

(a) Sketch the deformed body for the case $\phi L = \pi/2$.

(b) Compute the deformation gradient $\mathbf{F}$.

(c) Compute $\mathbf{C}$ at point $\mathbf{X} = (r_0/2, r_0/2, L/2)$ for $\phi = \pi/(2L)$.

6. Show that if λ_E is an eigenvalue of the Green–St. Venant strain tensor $\mathbf{E}$, and $\mathbf{n}_E$ is a corresponding eigenvector, then $1 + 2\lambda_E$ is an eigenvalue of $\mathbf{C}$ and $\mathbf{n}_E$ is also an eigenvector of $\mathbf{C}$ and that, thus, $\mathbf{E}$ and $\mathbf{C}$ have the same principal directions.

7. Show that $\mathbf{W}\boldsymbol{v} = \frac{1}{2}\boldsymbol{\omega} \times \boldsymbol{v}$.

8. Suppose that at a material point $\mathbf{X} \in \Omega_0$, the Green–St. Venant strain tensor is given by

$$\mathbf{E} = \sum_{1 \le i,j \le 3} E_{ij}\, \boldsymbol{e}_i \otimes \boldsymbol{e}_j,$$

where $\{e_i\}$ is an orthonormal basis in $\mathbb{R}^3$ and

$$[E_{ij}] = \begin{bmatrix} 3 & \sqrt{2} & 0 \\ \sqrt{2} & 2 & 0 \\ 0 & 0 & 1 \end{bmatrix}.$$

Determine:

(a) the principal directions of $\mathbf{E}$,

(b) the principal values of $\mathbf{E}$,

(c) the transformation $\mathbf{Q}$ that maps $\{e_i\}$ into the vectors defining the principal directions of $\mathbf{E}$,

(d) the principal invariants of $\mathbf{C} = 2\mathbf{E} + \mathbf{I}$.

9. Recall that an invariant of $\mathbf{C}$ is any real-valued function $\mu(\mathbf{C})$ such that $\mu(\mathbf{C}) = \mu(\mathbf{A}^{-1}\mathbf{C}\mathbf{A})$ for all invertible matrices $\mathbf{A}$. Show that tr $\mathbf{C}$, tr Cof $\mathbf{C}$, and det $\mathbf{C}$ are invariants of $\mathbf{C}$.

Exercise Set I.4

1. Reproduce the proof of Cauchy's Theorem for the existence of the stress tensor for the *two-dimensional* case ("plane stress") (to simplify geometric issues). Thus, for $\boldsymbol{\sigma}(\boldsymbol{n}) = \sigma_1(\boldsymbol{n})\boldsymbol{e}_1 + \sigma_2(\boldsymbol{n})\boldsymbol{e}_2$, show that $\exists\, \mathbf{T}$ such that $\boldsymbol{\sigma}(\boldsymbol{n}) = \mathbf{T}\boldsymbol{n}$.

2. The Cauchy stress tensor in a body $\mathcal{B}$ is

$$\mathbf{T}(\boldsymbol{x}, t) = T_{ij}(\boldsymbol{x}, t)\, \boldsymbol{e}_i \otimes \boldsymbol{e}_j,$$

where

$$T_{ij}(\boldsymbol{x}, t) = e^{10-10t} \begin{bmatrix} 10000x_1^2 - 7000x_1x_2 & 7000x_1x_3 & 2000x_3^2 \\ 7000x_1x_3 & 3000x_2^2 & 100x_1 \\ 2000x_3^2 & 100x_1 & 1000x_3^2 \end{bmatrix}.$$

(a) At point $\boldsymbol{x} = (1, 1, 1)$ at time $t = 1$, compute the stress vector $\boldsymbol{\sigma}(\boldsymbol{n})$ in the direction $\boldsymbol{n} = n_i\boldsymbol{e}_i$, $n_i = 1/\sqrt{3}$, $i = 1, 2, 3$.

(b) At $t = 1$, what is the total contact force on the plane surface $x_1 = 1, 0 \le x_2 \le 1, 0 \le x_3 \le 1$?

3. Let $\mathbf{T} = \mathbf{T}(\boldsymbol{x}, t)$ be the Cauchy stress at $\boldsymbol{x} \in \Omega_t$ at time t. If $\hat{n}$ is a direction (a unit vector) such that

$$\mathbf{T}\hat{n} = \sigma\hat{n} \qquad (\hat{n}^T \hat{n} = 1),$$

then (in analogy with principal values and directions of $\mathbf{C}$) σ is a *principal stress* and $\hat{n}$ is a *principal direction* of $\mathbf{T}$ (eigenvalues and eigenvectors of $\mathbf{T}$).

Continuing, let Γ be a plane through a point $\boldsymbol{x}$ with unit normal $\boldsymbol{n}$. The *normal stress* σ_n at $\boldsymbol{x}$ is

$$\sigma_n = (\boldsymbol{n} \cdot \mathbf{T}\boldsymbol{n}) \, \boldsymbol{n},$$

and the *shear stress* is

$$\sigma_t = \mathbf{T}\boldsymbol{n} - \sigma_n = \mathbf{T}\boldsymbol{n} - (\boldsymbol{n} \cdot \mathbf{T}\boldsymbol{n}) \, \boldsymbol{n}.$$

Show that if $\boldsymbol{n}$ were a principal direction of $\mathbf{T}$, then $\sigma_t = \mathbf{0}$.

4. Newton's Law of action and reaction asserts that for each $\boldsymbol{x} \in \overline{\Omega_t}$ and each t,
$$\boldsymbol{\sigma}(\boldsymbol{n}, \boldsymbol{x}, t) = -\boldsymbol{\sigma}(-\boldsymbol{n}, \boldsymbol{x}, t).$$

(a) Prove this law under the same assumptions as Cauchy's theorem.
    Hint:

    i. Let $\boldsymbol{n}$ be an arbitrary unit vector such that $\boldsymbol{n} \cdot \boldsymbol{e}_i > 0$. Consider the tetrahedron τ in Fig. E.2 with vertex $\boldsymbol{x}$ and distance d from $\boldsymbol{x}$ to face F. Show that the net force on τ is

$$\int_F \boldsymbol{\sigma}(\boldsymbol{n}, \boldsymbol{x}, t) \, dA + \sum_{i=1}^{3} \int_{F_i} \boldsymbol{\sigma}(-\boldsymbol{e}_i, \boldsymbol{x}, t) \, dA.$$

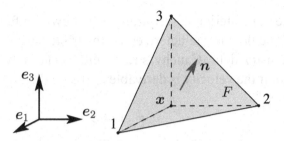

Figure E.2: Tetrahedral element.

ii. Noting that area(F)$\sim \mathcal{O}(d^2)$ and vol(τ)$\sim \mathcal{O}(d^3)$, apply the principle of balance of linear momentum (noting that the net body force $(\boldsymbol{f} - \rho d\boldsymbol{v}/dt)$ vanishes as $d \to 0$), leaving: $\boldsymbol{\sigma}(\boldsymbol{n}, \boldsymbol{x}, t) = -(\boldsymbol{n} \cdot \boldsymbol{e}_i)\boldsymbol{\sigma}(-\boldsymbol{e}_i, \boldsymbol{x}, t)$.

iii. Take $\boldsymbol{n} = \boldsymbol{e}_1$ (or $\boldsymbol{e}_2$ or $\boldsymbol{e}_3$) and conclude that since the basis $\{\boldsymbol{e}_j\}$ is arbitrary, the assertion stated in Problem 4 must hold.

(b) Complete the proof of Cauchy's Theorem: Show $\mathbf{T}(\boldsymbol{x}) = \mathbf{T}(\boldsymbol{x})^T$. (This proof makes use of (1) the principle of balance of angular momentum and (2) the equations of motion (div $\mathbf{T} + \mathbf{f} = \varrho \, d\boldsymbol{v}/dt$).)

Exercise Set I.5

1. Let Γ be a material surface associated with the reference configuration: $\Gamma \subset \partial\Omega_t$. Let $\boldsymbol{g}$ be an applied force per unit area acting on $\Gamma(\boldsymbol{g} = \boldsymbol{g}(\boldsymbol{x}, t), \boldsymbol{x} \in \Gamma)$. The "traction" boundary condition on Γ at each $\boldsymbol{x} \in \Gamma$ is

$$\mathbf{T}\boldsymbol{n} = \boldsymbol{g}.$$

Show that

$$\mathbf{F}\mathbf{S}\boldsymbol{n}_0 = \boldsymbol{g}_0, \qquad \text{on } \varphi^{-1}(\Gamma),$$

where $\boldsymbol{n}_0$ is the unit exterior normal to Γ_0 ($\Gamma = \varphi(\Gamma_0)$) and

$$\boldsymbol{g}_0(\boldsymbol{X}, t) = \det \mathbf{F}(\boldsymbol{X}) \| \mathbf{F}^{-T}(\boldsymbol{X})\boldsymbol{n}_0 \| \, \boldsymbol{g}(\boldsymbol{x}).$$

2. Consider an Eulerian description of the flow of a fluid in a region of $\mathbb{R}^3$. The flow is characterized by the triple $(v, \varrho, \mathbf{T}) = (\text{velocity field, density field, Cauchy stress field})$. The flow is said to be *potential* if the velocity is derivable as the gradient of a scalar field φ:

$$v = \text{grad } \varphi.$$

The body force field acting on the fluid is said to be *conservative* if there is also a potential U such that

$$\mathbf{f} = -\varrho \text{ grad } U.$$

The special case in which the stress $\mathbf{T}$ is of the form

$$\mathbf{T} = -p\,\mathbf{I},$$

where p is a scalar field and $\mathbf{I}$ is the unit tensor, is called a *pressure field* ($p =$ the static pressure or "hydrostatic" pressure).

(a) Show that for potential flow, a pressure field $\mathbf{T} = p\,\mathbf{I}$, and conservative body forces, the momentum equations imply that

$$\text{grad}\left(\frac{\partial\varphi}{\partial t} + \frac{1}{2}v \cdot v + U\right) + \frac{1}{\varrho}\text{ grad } p = \mathbf{0}.$$

This is Bernoulli's Equation for potential flow.
Hint: Show that

1) $\text{div } \mathbf{T} = -\text{grad } p,$
2) $dv/dt = \partial v/\partial t + \frac{1}{2}\text{ grad } v \cdot v.$

(b) If the motion is steady (i.e., if $\partial v/\partial t = 0$, $(v(x,t)$ is invariant with respect to t)) and $\partial\mathbf{f}/\partial t = 0$, so $\partial U/\partial t = 0$, then the equations of part (a) reduce to

$$v \cdot \text{grad}\left(\frac{1}{2}v \cdot v + U\right) + \frac{1}{\varrho}v \cdot \text{grad } v = 0$$

(notice that $v \cdot dv/dt = v \cdot \text{grad}(\frac{1}{2}v \cdot v)$).

3. Let $u = u(X, t) = \varphi(X, t) - X$ be the displacement field. Define the quantity

$$\psi = \frac{1}{2} \int_{\Omega_0} \varrho_0 u \cdot u \, dX.$$

Show that (if $\mathbf{f}_0 = 0$),

$$\ddot{\psi} = \int_{\Omega_0} \varrho_0 \dot{u} \cdot \dot{u} \, dX - \int_{\Omega_0} \mathbf{P} : \nabla u \, dX + \int_{\partial\Omega_0} u \cdot \mathbf{P} \mathbf{n}_0 \, dA_0.$$

4. A cylindrical rubber plug 1 cm in diameter and 1 cm long (see Fig. E.3) is glued to a rigid foundation. Then it is pulled by external forces so that the flat cylindrical upper face $\Gamma_0 = \{(X_1, X_2, X_3) : X_3 = 1, \ (X_1^2 + X_2^2)^{1/2} \leq 1/2\}$ is squeezed to a flat circular face Γ of diameter 1/4 cm with normal $n = e_2$, as shown at position $x = x^*$. Suppose that the stress vector at $x = x^*$ is uniform and normal to Γ:

$$\sigma(n, x^*) = \sigma(e_2, x^*) = 1000 \, e_2 \ \text{kg/cm}^2, \qquad \forall x^* \in \Gamma.$$

Suppose that the corresponding Piola-Kirchhoff stress $\mathbf{p}_0 = \mathbf{P} n_0$ is uniform on Γ_0 and normal to Γ_0.

(a) Determine the Piola–Kirchhoff stress vector $\mathbf{p}_0 = \mathbf{P} n_0$ on Γ_0.

(b) Determine one possible tensor Cof $\mathbf{F}(X_1, X_2, 1)$ for this situation.

Exercise Set I.6

1. Consider the small deformations and heating of a thermoelastic solid constructed of a material characterized by the following constitutive equations:

Free energy: $\varrho_0 \psi_0 = \frac{1}{2}\lambda(\text{tr } e)^2 + \mu e : e + c(\text{tr } e)\theta + \frac{c_0}{2}\theta^2,$

Heat flux: $\mathbf{q}_0 = k\nabla\theta,$

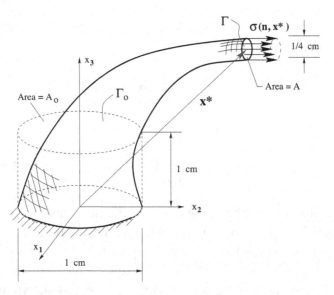

Figure E.3: Illustrative sketch of the rubber plug.

where

$$e = \frac{1}{2}(\nabla u + \nabla u^T) = \text{the ``infinitesimal" strain tensor } (\approx \mathbf{E}),$$

$u = $ the displacement field,

$\theta = $ the temperature field,

$\lambda, \mu, c, c_0, k = $ material constants.

A body $\mathcal{B}$ is constructed of such a material and is subjected to body forces $\mathbf{f}_0$ and to surface contact forces g on a portion Γ_g of its boundary $\Gamma_g \subset \partial\Omega_0$. On the remainder of its boundary, $\Gamma_u = \partial\Omega_0 \backslash \Gamma_g$, the displacements u are prescribed as zero ($u = 0$ on Γ_u). The mass density of the body is ϱ_0, and, in its reference configuration at time $t = 0$, we have $u(x, 0) = u_0(x)$, $\partial u(x, 0)/\partial t = v_0(x)$, $x \in \Omega_0$, where u_0 and v_0 are given functions. A portion Γ_q of the boundary is heated, resulting in a prescribed heat flux $h = \mathbf{q} \cdot \mathbf{n}$, and the complementary boundary, $\Gamma_\theta = \partial\Omega_0\backslash\Gamma_q$ is subjected to a prescribed temperature $\theta(\boldsymbol{x},t) = \tau(\boldsymbol{x},t)$, $\boldsymbol{x} \in \Gamma_\theta$ (e.g. is immersed in ice water).

Develop a mathematical model of this physical system (a set of partial differential equations, boundary and initial conditions) describing the dynamic, thermomechanical behavior of a thermoelastic solid.

EXERCISES FOR PART II

Exercise Set II.1

1. Ten football teams in a college conference had the following record last season: Three teams 7 and 4 (7 wins, 4 losses), four teams 8 and 3, two teams 9 and 2, and one team 11 and 0. (Obviously, some of these included out-of-conference games.) The total number of teams in the conference:

$$N = \sum_{j=0}^{\infty} N(j) = 10,$$

where $N(j)$ is the number of teams with j wins. This set of teams is the sample set Ω.

(a) Show (give a brief one-line argument) that the probability that a team selected randomly has j wins is

$$\mathbb{P}(j) = \frac{N(j)}{N}.$$

(b) Show another (obvious) property,

$$\sum_{j=0}^{\infty} \mathbb{P}(j) = 1 \quad \left(= \mathbb{P}(\Omega) \right).$$

(c) Demonstrate by a full calculation that

$$\langle j \rangle = \sum_{j=0}^{\infty} j \, \mathbb{P}(j).$$

(Show that $\langle j \rangle$ is precisely the sum on the right-hand side of this equality.)

(d) Compute the variance and standard deviation,

$$\sigma^2 = \langle j^2 \rangle - \langle j \rangle^2 = \sum_{j=0}^{\infty} \left(j - \langle j \rangle \right)^2 \mathbb{P}(j).$$

2. The Gaussian probability density function in one dimension is of the form

$$\rho(x) = C e^{-\frac{\alpha}{2}(x - x_0)^2},$$

where x_0 is a point on the real line and C and α are constants.

(a) Determine C.

(b) Determine $\langle x \rangle, \langle x^2 \rangle$.

(c) Determine σ.

(d) Sketch a graph of $\rho(x)$.

3. Let

$$\Psi(x) = \begin{cases} A\,(1 + x), & -1 \le x \le 0, \\ A\,(1 - x), & 0 \le x \le 1, \\ 0, & x \notin [-1, 1]. \end{cases}$$

(a) Determine A.

(b) Determine $\langle x \rangle, \langle x^2 \rangle$.

(c) Determine σ_x.

(d) Determine $\langle p \rangle$.

4. The state of a quantum system is given by

$$\Psi(x, t) = \alpha e^{-\Lambda},$$

where

$$\Lambda = \beta \hbar^{-1} \left(m x^2 + i \gamma t \right),$$

and in which α, β, and γ are constants.

(a) Find the potential $V(x)$ of this system.

(b) Calculate the expected values of x, x^2, p, and p^2.

(c) What is α?

(d) Calculate σ_x and σ_p. Are these consistent with the Heisenberg principle?

Exercise Set II.2

1. The proof of (11.27) follows from several algebraic steps and the Cauchy–Schwarz inequality. Let

$$u = \left(\tilde{Q} - \langle Q \rangle \right) \Psi \quad \text{and} \quad v = \left(\tilde{M} - \langle M \rangle \right) \Psi.$$

(a) Then verify that

$$\begin{aligned}
\sigma_Q^2 &= \langle Q^2 \rangle - \langle Q \rangle^2 \\
&= \langle \Psi, \tilde{Q}(\tilde{Q}\Psi) \rangle - \langle \Psi, \langle Q \rangle^2 \Psi \rangle \\
&= \langle \tilde{Q}\Psi, \tilde{Q}\Psi \rangle - \langle \langle Q \rangle \Psi, \langle Q \rangle \Psi \rangle \\
&= \langle (\tilde{Q} - \langle Q \rangle) \Psi, (\tilde{Q} - \langle Q \rangle) \Psi \rangle \qquad \text{(why?)} \\
&= \langle u, u \rangle \\
&= \|u\|^2,
\end{aligned}$$

and similarly, $\sigma_M^2 = \|v\|^2$. Thus, by Cauchy–Schwarz,

$$\sigma_Q^2 \, \sigma_M^2 = \|u\|^2 \, \|v\|^2 \geq |\langle u, v \rangle|^2.$$

(b) Show that

$$\langle u, v \rangle = \langle QM \rangle - \langle Q \rangle \langle M \rangle.$$

Hint: Show that

$$\langle u, v \rangle = \langle \Psi, \tilde{Q}\tilde{M}\Psi \rangle - \langle M \rangle \langle \Psi, \tilde{Q}\Psi \rangle - \langle Q \rangle \langle \Psi, \tilde{M}\Psi \rangle$$
$$+ \langle Q \rangle \langle M \rangle \langle \Psi, \Psi \rangle$$
$$= \langle QM \rangle - \langle Q \rangle \langle M \rangle.$$

(c) The number $\langle u, v \rangle$ is complex. From the fact that any complex number z satisfies

$$|z|^2 \geq \left(\frac{1}{2i}(z - z^*) \right)^2,$$

where z^* is the complex conjugate of z, show that

$$\sigma_Q^2 \sigma_M^2 \geq \left(\frac{1}{2i} \left(\langle u, v \rangle - \langle v, u \rangle \right) \right)^2.$$

(d) From (b) and (c), show that (11.27) is true.

2. A classical textbook example (see, e.g., Griffiths [21, p. 25]) of the time-independent Schrödinger equation for a single particle moving along the x axis is the problem of the infinite square well for which the potential V is of the form

$$V(x) = \begin{cases} 0, & \text{if } 0 \leq x \leq a, \\ \infty, & \text{otherwise.} \end{cases}$$

The particle is confined to this "well", so $\psi(x) = 0$ for $x < 0$ or $x > a$; inside the well, $V(x) \equiv 0$.

a) Show that Schrödinger's equation reduces to

$$\frac{d^2\psi}{dx^2} + k^2 \psi = 0, \qquad \psi(0) = \psi(a) = 0,$$

with $k^2 = 2mE/\hbar^2$.

b) Show that the possible values of the energy are

$$E_n = \frac{n^2\pi^2\hbar^2}{2ma^2}, \qquad n = 1, 2, \dots.$$

c) Show that the wave function is a superposition of solutions,

$$\psi_n(x) = \sqrt{\frac{2}{a}} \sin\left(\frac{n\pi x}{a}\right), \qquad 0 \le x \le a,$$

which are orthonormal in $L^2(0, a)$.

d) Given that the functions ψ_n in (c) form a complete orthonormal basis for $L^2(0, a)$, develop a Fourier series representation of the function $f(x) = x$ as an infinite sum of these functions.

e) Let

$$Q(x, p) \,\Psi = d^2\Psi(x)/dx^2, \qquad 0 \le x \le a.$$

Is Q an observable?

3. Consider a quantum system consisting of a single particle in a straight line with position x and momentum p. Define the following operators:

$$\begin{aligned} A\varphi &= x\varphi & (A = \text{``}x\text{''}), \\ B\varphi &= \partial\varphi/\partial x & (B = \partial/\partial x). \end{aligned}$$

Do A and B commute?

4. Let

$$Q = -\frac{\hbar^2}{2m}\frac{\partial^2}{\partial x^2} \quad \text{and} \quad M = -i\hbar\frac{\partial}{\partial x} \quad (= \hat{p}).$$

a) Show that these operators commute.

b) Show that e^{ikx} is a simultaneous eigenfunction of these operators and, indeed, $p^2/2m = E$.

5. Consider the eigenfunctions of the momentum operator $p_x = -i\hbar\, \partial/\partial x$:

$$-i\hbar\frac{\partial\psi_m}{\partial x} = \lambda_p\psi_m.$$

(a) Verify that the eigenfunctions are

$$\psi_m(x) = A\, e^{i\lambda_p x/\hbar},$$

where A is a complex constant.

(b) Verify that the momentum eigenfunctions are such that $\psi^*\psi = A^*A$, a constant independent of x (meaning ψ is a plane wave) and that, therefore, x is indeterminate, in keeping with Heisenberg's principle.

6. The Dirac delta "function," $\delta(x)$, is really not a function in the usual sense. It is a distribution (a continuous linear functional on the space $\mathcal{D}$ of test functions φ) such that

$$\langle\delta,\varphi\rangle = \varphi(0) \qquad \forall\varphi\in\mathcal{D},$$

where $\mathcal{D}$ is the space of infinitely differentiable functions defined on $\mathbb{R}$ with compact support ($\varphi(\pm\infty) = 0$), and $\langle\cdot,\cdot\rangle$ denotes pairing on the space of distributions $\mathcal{D}'$ (dual to $\mathcal{D}$) and $\mathcal{D}$, and $\mathcal{D}$ is equipped with a special topological structure (see [6]). It is customary to use a symbolic notation,

$$\langle\delta,\varphi\rangle = \text{``}\int_{-\infty}^{\infty}\delta(x)\,\varphi(x)\,dx\text{''} = \varphi(0) \qquad \forall\varphi\in\mathcal{D}$$

(or "$\int_{-\infty}^{\infty}\delta(x)\,dx$" $= 1$). A translation of δ has the property

$$\langle\delta(x-a),\varphi\rangle = \varphi(a) \qquad \forall\varphi\in\mathcal{D}.$$

Now let us consider the potential (cf. Griffiths [21, Sec. 2.5]):

$$V(x) = -\sigma\,\delta(x), \qquad \sigma\in\mathbb{R}^+.$$

The corresponding Schrödinger equation is the (distributional) differential equation

$$-\frac{\hbar}{2m}\frac{d^2\psi}{dx^2} - \sigma\,\delta(x)\,\psi = E\psi.$$

(a) Show that for $x < 0$, $\psi(x) = Ae^{kx}$, $k = \sqrt{-2mE}/\hbar$ and for $x > 0$, $\psi(x) = Be^{-kx}$, A and B constants.

(b) Owing to the fact that ψ must be continuous at $x = 0$, show that

$$\psi(x) = \begin{cases} A\,e^{kx}, & x \le 0, \\ A\,e^{-kx}, & x > 0. \end{cases}$$

(c) From the condition

$$0 = \lim_{\epsilon\to 0}\left(-\frac{\hbar}{2m}\int_{-\epsilon}^{\epsilon}\frac{d\psi'}{dx}\,dx + \int_{-\epsilon}^{\epsilon}V(x)\,\psi(x)\,dx\right.$$
$$\left. - E\int_{-\epsilon}^{\epsilon}\psi(x)\,dx\right),$$

show that

$$\Delta\psi' = \psi'|_{0+} - \psi'|_{0-} = (-2m\sigma/\hbar^2)\,\psi(0),$$

and that therefore $k = m\sigma/\hbar^2$.

(d) Since ψ must be normalized ($\int_{-\infty}^{\infty}|\psi|^2\,dx = 1$), show that $A = \sqrt{k}$ and, hence,

$$\psi(x) = \sqrt{\frac{m\sigma}{\hbar^2}}\,e^{-m\sigma|x|/\hbar^2} \quad\text{and}\quad E = -m\sigma/2\hbar^2.$$

7. Which of the following operators are Hermitian?

(a) e^{ix} (1D);

(b) $\dfrac{d^2}{dx^2}$ (1D);

(c) $x \dfrac{d}{dx}$ (1D).

8. Ehrenfest's theorem relates the time derivative of the expected value $\langle Q \rangle$ of an operator $\tilde{Q}$ to the commutator $[\tilde{Q}, H]$ of the operator of the Hamiltonian of the system and the expected value of $\partial Q / \partial t$ as follows:

$$\frac{d \langle Q \rangle}{dt} = \frac{1}{i\hbar} \langle [\tilde{Q}, H] \rangle + \left\langle \frac{\partial Q}{\partial t} \right\rangle.$$

Derive the following intermediate results:

(a) $\dfrac{d \langle Q \rangle}{dt} = \displaystyle\int_{\mathbb{R}^3} \frac{\partial \Psi^*}{\partial t} \tilde{Q} \Psi \, d^3x + \left\langle \frac{\partial Q}{\partial t} \right\rangle + \int_{\mathbb{R}^3} \Psi^* \tilde{Q} \frac{\partial \Psi}{\partial t} \, d^3x,$

(b) $$\frac{\partial \Psi^*}{\partial t} = -\frac{1}{i\hbar} H \Psi^*.$$

Note that

$$\langle [\tilde{Q}, H] \rangle \equiv \int_{\mathbb{R}^3} \Psi^* [\tilde{Q}, H] \Psi \, d^3x.$$

c) Use (a) and (b) to complete the proof of Ehrenfest's theorem.

Exercise Set II.3

1. This exercise is designed to carry through the classical method of separation of variables for the solution of partial differential equations. The problem is the two-dimensional *particle in a box* (which generalized the 1D problem of Exercise II.2.2 in the previous chapter). The physical situation is that of a single particle in a square box $\overline{\Omega} = [0, a] \times [0, b]$ in the xy plane in a quantum system

for which the potential $V = V(x, y)$ is

$$V(x, y) = \begin{cases} 0 & \text{for } 0 \le x \le a, \ 0 \le y \le b, \\ \infty & \text{otherwise.} \end{cases}$$

The Hamiltonian is thus

$$H = \begin{cases} (-\hbar^2/2m) \, \Delta & \text{in the box } ((x, y) \in \overline{\Omega}), \\ +\infty & \text{outside the box,} \end{cases}$$

where Δ is the two-dimensional Laplacian,

$$\Delta = \frac{\partial^2}{\partial x^2} + \frac{\partial^2}{\partial y^2}.$$

Thus, the wave function $\Psi = 0$ outside $\overline{\Omega}$. Considering the time-independent Schrödinger equation, we wish to find $\psi = \psi(x, y)$ satisfying

$$H\psi = E\psi \qquad \text{in } \overline{\Omega},$$
$$\psi = 0 \qquad \text{on } \partial\overline{\Omega}.$$

(a) Use the standard trick of the method of separation of variables: Assume ψ is a product of a function $X(x)$ and a function $Y(y)$: $\psi(x, y) = X(x) Y(y)$, and derive two ordinary differential equations, one for X and one for Y under the assumption that E is the sum, $E = e_X + e_Y$, $e_X = $ constant, $e_Y = $ constant. Show that the solutions are of the form

$$\psi_{nn'}(x, y) = \sqrt{\frac{4}{ab}} \sin \frac{n\pi x}{a} \sin \frac{n'\pi y}{b}.$$

(b) Show that the energy levels are given by

$$E = \frac{\hbar^2 \pi^2}{2m} \left(\frac{n^2}{a^2} + \frac{n'^2}{b^2} \right).$$

(c) For the case of a square box, $(a = b)$, determine the energy levels $E = (\hbar^2 \pi^2 / 2ma^2)(n^2 + n'^2)$ for values of n and n' up to 4. What is the lowest energy level of the system?

2. This exercise concerns a well-known argument for using the nucleus of the hydrogen atom as the origin of the coordinate system as opposed to the center of mass. The situation is this: The quantum system consists of two particles, particle 1 of mass M (corresponding, e.g., to the nucleus) and particle 2 of mass m (e.g. the electron), with origin at a point O. The particles are located at positions $\mathbf{r}_1$ and $\mathbf{r}_2$, respectively. The center of mass is located at $\mathbf{R} = (\mathbf{r}_1 M + \mathbf{r}_2 m)/(M + m)$ and the vector connecting 1 to 2 is $\mathbf{r} = \mathbf{r}_1 - \mathbf{r}_2$. The goal is to rederive Schrödinger's equations with a change of variables from $(\mathbf{r}_1, \mathbf{r}_2)$ to $(\mathbf{R}, \mathbf{r})$.

(a) Show that

$$\mathbf{r}_1 = \mathbf{R} + \frac{m^*}{M}\mathbf{r} \quad \text{and} \quad \mathbf{r}_2 = \mathbf{R} - \frac{m^*}{m}\mathbf{r},$$

where $m^* = Mm/(M + m)$.

(b) Using the change of variables, show that

$$\nabla_{\mathbf{r}_1} = \frac{m^*}{m}\nabla_{\mathbf{R}} + \nabla_{\mathbf{r}} \quad \text{and} \quad \nabla_{\mathbf{r}_2} = \frac{m^*}{M}\nabla_{\mathbf{R}} - \nabla_{\mathbf{r}}.$$

(c) Show that, in these new variables, the time-independent Schrödinger equation is

$$\left(-\frac{\hbar^2}{2(M+m)}\Delta_{\mathbf{R}} - \frac{\hbar^2}{2m^*}\Delta_{\mathbf{r}} + V(\mathbf{r})\right)\psi = E\,\psi.$$

(d) Now if $M \gg m$ so that $M+m \approx M$ and $1/m \gg 1/(M+m)$, write down the resulting approximate Schrödinger equation involving only $\mathbf{r}$.

(e) Suppose now that the wave function is separable: $\psi(\mathbf{r}, \mathbf{R}) = \psi(\mathbf{r})\,\chi(\mathbf{R})$. Show that χ satisfies the one-particle Schrödinger equation with mass $M + m$, with potential $V_{\mathbf{R}} = 0$ and energy $E_{\mathbf{R}}$, while ψ satisfies the one-particle Schrödinger equation with mass m^*, potential $V(\mathbf{r})$, and energy $E_{\mathbf{r}}$, with the total energy $E = E_{\mathbf{R}} + E_{\mathbf{r}}$.

Exercise Set II.4

1. Prove that the following correspondence principle holds for the principle of balance of angular momentum:

$$\frac{d \langle \mathbf{L} \rangle}{dt} = \langle \mathbf{q} \times \mathbf{F} \rangle, \qquad \mathbf{F} = -\frac{dV}{d\mathbf{q}}.$$

2. Show that the eigenvalues of the spin operator S_1 (for an electron) are $\pm \hbar/2$. What are the corresponding eigenvectors?

3. Confirm that S_1, S_2, S_3, and S^2 (for electrons) represent observables but S_+ and S_- do not.

4. Ignoring normalization, determine which (if any) of the following wave functions are valid for the helium atom.

a) $\varphi_{1s}(1)\, \varphi_{2s}(2) \begin{vmatrix} \alpha_1 & \beta_1 \\ \alpha_2 & \beta_2 \end{vmatrix}$

b) $\left(\varphi_{1s}(1)\, \varphi_{2s}(2) - \varphi_{2s}(1)\, \varphi_{1s}(2) \right) \begin{vmatrix} \alpha_1 & \beta_1 \\ \alpha_2 & \beta_2 \end{vmatrix}$

EXERCISES FOR PART III

Exercise Set III.1

1. The differential of the internal energy $U = U(S, V, N)$ is

$$dU = TdS - p\,dV + \mu dN,$$

with T, S, p, V, μ, N the temperature, entropy, pressure, volume, chemical potential, and number of particles, respectively.

(a) Let

$$
\begin{aligned}
h &= h(S, p, N) &&= U + pV = \text{ the enthalpy,} \\
A &= A(T, V, N) &&= U - TS = \text{ the Helmholtz free energy,} \\
G &= G(T, p, N) &&= U + pV - TS = \text{ the Gibbs free energy,}
\end{aligned}
$$

derive the differentials

$$
\begin{aligned}
dh(S, p, N) &= TdS + V\,dp + \mu\,dN, \\
dA(T, V, N) &= -SdT - p\,dV + \mu\,dN, \\
dG(T, p, N) &= -SdT + V\,dp + \mu\,dN.
\end{aligned}
$$

b) Let $U = U(S, V, N) = TS - pV + \mu N$, derive the relations

$$
\begin{aligned}
h &= TS + \mu N, \\
A &= -pV + \mu N, \\
G &= \mu N,
\end{aligned}
$$

and, in addition, derive the equation

$$SdT - Vdp + Nd\mu = 0.$$

c) **Maxwell Relations**: The Maxwell relations are relationships between derivatives of various thermodynamic potentials that can be derived in a straightforward manner.

(i) Derive the Maxwell relations

$$-\left(\frac{\partial p}{\partial S}\right)_{V,N} = \left(\frac{\partial T}{\partial V}\right)_{S,N} = \frac{\partial^2 U}{\partial V \partial S},$$

$$\left(\frac{\partial T}{\partial N}\right)_{S,V} = \left(\frac{\partial \mu}{\partial S}\right)_{V,N} = \frac{\partial^2 U}{\partial N \partial S},$$

$$-\left(\frac{\partial p}{\partial N}\right)_{S,V} = \left(\frac{\partial \mu}{\partial V}\right)_{S,N} = \frac{\partial^2 U}{\partial V \partial N}.$$

Hint: The proof of these relations follows from elementary use of the definitions of the total differential $dU = TdS - pdV + \mu dN$ and the fact that the order of differentiation can be interchanged (e.g., $\partial^2 U/\partial S\partial V = \partial^2 U/\partial V\partial S$).

(ii) Derive similar Maxwell relations from the thermodynamic potentials h, A, and G.

2. The partition function of an ideal gas in the canonical ensemble is (recall (17.48))

$$\Omega_{\text{ideal}} = \frac{V^N}{N!h^{3N}}(2\pi mkT)^{3N/2},$$

where V is the volume occupied by the N particles, h is Planck's constant, m is the mass of each particle, T is the temperature, and k is the Boltzmann constant.

(a) Use the relation between the Helmholtz free energy A and the partition function, for the case of an ideal gas, i.e., $A =$

$-kT \ln \Omega_{\text{ideal}}$ (cf. (17.26)), to obtain

$$A = -kTN \ln \left(\frac{eV}{N} \right) - \frac{3N}{2} kTN \ln \left(\frac{2\pi mkT}{h^2} \right),$$

with e the base of the natural logarithm.

Hint: Use Stirling's approximation, in the more precise form: $N! \approx \sqrt{2\pi N} \left(\frac{N}{e} \right)^N$. Further, for $N \gg 1$ we can assume terms of order $\mathcal{O}(\ln(N))$ to be negligible in comparison to terms of order $\mathcal{O}(N)$.

(b) Use the relation

$$\left(\frac{\partial A}{\partial V} \right)_{T,N} = -p$$

to obtain the equation for an ideal gas, i.e.,

$$pV = NkT.$$

(c) The Helmholtz free energy is related to the internal energy by the relation $A = U - TS$. Use the relation

$$\left(\frac{\partial A}{\partial T} \right)_{V,N} = -S$$

to obtain the equation

$$U = \frac{\partial}{\partial \beta} (\beta A),$$

with $\beta = 1/kT$. Then, derive the internal energy of an ideal gas:

$$U = \frac{3}{2} NkT.$$

3. Recall that in Chapter 6 the change in entropy in bringing a thermodynamic system from state 1 to state 2 satisfies

$$\Delta S = S(\mathcal{S}, t_2) - S(\mathcal{S}, t_1) \geq \frac{1}{T} \delta q$$

for an isothermal process, where δq denotes the heat absorbed by moving from 1 to 2:

$$\delta q = Q = \int_1^2 dq.$$

The equality $\Delta S = \delta q / T$ holds for a reversible process.

(a) Show that this relation implies that the heat absorbed in a reversible process, δq_{rev}, is greater than or equal to δq_{irrev}, that absorbed in an irreversible process, in the case of a simple one-component system for which $dE = -p\,dV + T\,dS$ and $dA = dE - T\,dS - S\,dT$.

(b) Show that for such a system, the free energy A is a minimum for constant T and V (i.e., $\Delta A \leq 0$).

4. In the case of multi-component systems, in which the number N_j of moles of constituent j of the system can change, the work done due to a change dN_j in N_j is written $\mu_j dN_j$, $j = 1, 2, \ldots, M$ (M constituents), where μ_j is the *chemical potential* of the component j. The first law of thermodynamics (recall (17.1)) can be written, in this case, as the equality

$$\frac{d}{dt}(\kappa + U) = \frac{dE}{dt} = \mathbf{S} : \dot{\mathbf{E}} + Q + \sum_{j=1}^{M} \mu_j \frac{dN_j}{dt},$$

where (for instance) $\mathbf{S}$ is the second Piola–Kirchhoff stress and $\dot{\mathbf{E}}$ is the time rate of change of the Green–St. Venant strain. In classical equilibrium statistical mechanics, it is customary to merely consider the stress power produced by the pressure p acting on a change in volume dV, so $\mathbf{S} : d\mathbf{E}/dt$ is replaced by $-p\,dV/dt$. The heating per unit time is set equal to $T\,dS/dt$. Thus, we have

$$dE = -p\,dV + T\,dS + \sum_{j=1}^{N} \mu_j dN_j.$$

Using the appropriate Legendre transformations, derive the corresponding equations for differentials of the enthalpy $h = h(S, p)$, the Helmholtz free energy $A = A(T, V)$, and the Gibbs free energy $G = G(T, p)$:

$$dh = V dp + T dS + \sum_{j=1}^{M} \mu_j dN_j,$$

$$dA = -p \, dV - S dT + \sum_{j=1}^{M} \mu_j dN_j,$$

$$dG = V dp - S dT + \sum_{j=1}^{M} \mu_j dN_j.$$

5. The specific heats at constant volume V and number N of constituents, C_V, and at constant pressure p and N, C_p, are linear functions of the temperature T:

$$C_V = T \left(\frac{\partial S}{\partial T} \right)_{V,N}, \qquad C_p = T \left(\frac{\partial S}{\partial T} \right)_{p,N}.$$

a) Show that

$$\left(\frac{\partial C_V}{\partial V} \right)_{T,N} = T \left(\frac{\partial^2 p}{\partial T^2} \right)_{V,N}.$$

b) Show that

$$C_p - C_V = T \left(\frac{\partial p}{\partial T} \right)_{V,N} \left(\frac{\partial V}{\partial T} \right)_{p,N}$$

$$= -T \left(\frac{\partial p}{\partial V} \right)_{T,N} \left[\left(\frac{\partial V}{\partial T} \right)_{p,N} \right]^2.$$

6. Recall that an *extensive* property of a system is a macroscopic property that depends, generally speaking, linearly on the size of the system; e.g., if two systems A and B, with energies E_A and E_B respectively are combined to make a composite system, the energy of the composite is the sum $E_A + E_B$. Extensive properties are thus characterized by first-order homogeneous functions, i.e., functions $f = f(x)$ such that $f(\lambda x) = \lambda f(x)$, $\lambda \in \mathbb{R}$.

(a) Prove *Euler's theorem for first-order homogeneous functions*: if $f(\lambda\mathbf{x}) = \lambda f(\mathbf{x})$, $\mathbf{x} = (x_1, x_2, \ldots, x_n)$ and f is differentiable, then

$$f(\mathbf{x}) = \nabla f(\mathbf{x}) \cdot \mathbf{x} = \sum_{i=1}^{n} \frac{\partial f(\mathbf{x})}{\partial x_i} x_i.$$

(b) Suppose that the total energy E is a function of the entropy S and a kinematical variable $\mathbf{u}$ so that $E = E(\lambda S, \lambda\mathbf{u})$, with $(\partial E/\partial\mathbf{u}) \cdot \mathbf{u} = -p\,dV + \sum_{j=1}^{M} \mu_j dN_j$. Show that Euler's theorem implies that

$$E = TS - p\,dV + \sum_{j=1}^{M} \mu_j dN_j.$$

(c) Derive the *Gibbs–Duhem equation*: at constant T and p,

$$\sum_{j=1}^{M} N_j d\mu_j = 0.$$

7. In the grand canonical ensemble, we encounter $\mathcal{N}$ identical systems mutually sharing $N\mathcal{N}$ particles, N particles per system, and a total energy $\mathcal{N}E$. We denote $n_{i,j}$ the number of systems that have N_i particles and an amount of energy E_j, so that

$$\sum_{i,j} n_{i,j} = \mathcal{N}, \tag{1a}$$

$$\sum_{i,j} n_{i,j} N_i = \mathcal{N}N, \tag{1b}$$

$$\sum_{i,j} n_{i,j} E_j = \mathcal{N}E. \tag{1c}$$

The multiplicity of distributions of particles and energy among the ensemble is

$$\Gamma(n_{i,j}) = \frac{\mathcal{N}!}{\Pi_{i,j}(n_{i,j}!)}.$$

Let $n_{i,j}^*$ denote the most probable distribution that maximizes Γ while satisfying the constraints (1a), (1b), and (1c) above. Show that

$$\frac{n_{i,j}^*}{N} = \frac{\exp(-\alpha N_i - \beta E_j)}{\sum_{i,j} \exp(-\alpha N_i - \beta E_j)},$$

where $\beta = 1/kT$ and $\alpha = -\mu/kT$.

8. Continuing the study of the grand canonical ensemble initiated in the previous exercise, let

$$\omega \overset{\text{def}}{=} \exp(-\alpha N_i - \beta E_j)$$

and

$$\Omega = \sum_{i,j} \omega_{i,j}.$$

Define

$$\langle N \rangle = \sum_{i,j} N_i \omega_{i,j}/\Omega, \qquad \langle E \rangle = \sum_{i,j} E_j \omega_{i,j}/\Omega.$$

(a) Show that

$$\langle N \rangle = -\frac{\partial}{\partial \alpha} q, \qquad \langle E \rangle = -\frac{\partial}{\partial \beta} q,$$

where

$$q = \ln\left(\sum_{i,j} \omega_{i,j} \right).$$

(b) Show that

$$dq = -\langle N \rangle d\alpha - \langle E \rangle d\beta - \frac{\beta}{N} \sum_{i,j} \langle n_{i,j} \rangle dE_j.$$

(c) Show that the first law of thermodynamics leads to the relation

$$q + \alpha \langle N \rangle + \beta \langle E \rangle = S/k,$$

with $\beta = 1/kT$ and $\alpha = -\mu/kT$.

(d) Noting that $\mu\langle N \rangle = G$, the Gibbs free energy $= \langle E \rangle - TS + pV$, show that

$$q = pV/kT.$$

Exercise Set III.2

1. Show that $\int_{\mathbb{R}^3} \mathbf{u} f \, dw = 0$.

2. Show that if (17.69) and (17.70) hold, then, in the spatial (Eulerian) description, we obtain

$$\varrho \frac{\partial e}{\partial t} = \mathbf{T} : \operatorname{grad} \mathbf{v} - \operatorname{div} \mathbf{q}.$$

BIBLIOGRAPHY

Part I

[1] BATRA, R., **Elements of Continuum Mechanics**, AIAA Education Series, Reston, VA, 2006.

[2] CIARLET, P.G., **Mathematical Elasticity, Volume 1: Three-Dimensioned Elasticity**, North-Holland, Amsterdam, 1988.

[3] ERINGEN, A.C., **Nonlinear Theory of Continuous Media**, McGraw-Hill, New York, 1962.

[4] GURTIN, M.E., **An Introduction to Continuum Mechanics**, Academic Press, New York, 1981.

[5] ODEN, J.T., **Finite Elements of Nonlinear Continua**, McGraw-Hill, New York, 1972, (Dover ed., 2006).

[6] ODEN, J.T. AND DEMKOWICZ, L.F., **Applied Functional Analysis**, Second Edition, CRC Press, Taylor and Francis Group, Boca Raton, FL, 2010.

[7] TRUESDELL, C. AND NOLL, W., **The Non-Linear Field Theories of Mechanics**, Handbuch der Physik, Vol. III/3, Springer, Berlin, 1965.

[8]   TRUESDELL, C. AND TOUPIN, R., **The Classical Field Theories**, Handbuch der Physik, Vol. III/1, Springer, Berlin, 1960.

Part II

[9]   ALCOCK, N.W., **Bonding and Structure: Structural Principles in Inorganic and Organic Chemistry**, Ellis Horwood, New York and London, 1990.

[10] BLOCH, F., *"Uber die Quantenmechanik der Elektron in Kristallgittern,"* **Z. Phys.**, 52, pp. 522–600, 1928.

[11] BORN, M., **Atomic Physics**, Dover, New York, Eighth Edition, Revised by R. J. Blin-Stoyle and J. M. Radcliffe, 1989.

[12] CANCES, E., *"SCF algorithms for HF electronic calculations,"* in M. DEFRANCESCHI AND C. LEBRIS (eds.), **Mathematical Models and Methods for Ab Initio Quantum Chemistry**, Lecture Notes in Chemistry, Springer-Verlag, Berlin, pp. 17–33, 2000.

[13] CARTMELL, E. AND FOWLES, G.W.A., **Valency and Molecular Structures**, Butterworths, London, 1977.

[14] DEFRANCESCHI, M. AND LEBRIS, C., (eds.), **Mathematical Models and Methods in Ab Initio Quantum Chemistry**, Springer-Verlag, Berlin, 2000.

[15] DREIZLER, R.M. AND GROSS, E.K.U., **Density Functional Theory**, Springer-Verlag, Berlin, 1993.

[16] FEYNMAN, R.P., **Phys. Rev.**, 56, p. 340, 1939.

[17] FOCK, V., *"Naherungsmethode zur lösung des quantum mechanischen Mehrkörperproblems,"* **Z. Phys.**, 61, pp. 126–148, 1930.

[18] GIL, V., **Orbitals in Chemistry**, Cambridge University Press, New York, 2000.

[19] GILL, P.M.W., ADAMSON, R.D. AND POPLE, J.A., *"Coulomb-attenuated exchange energy density functionals,"* **Mol. Phys.**, 88(4), pp. 1005–1010, 1996.

[20] GILLEPSIE, R.J. AND POPELIER, P.L., **Chemical Bonding and Molecular Geometry**, Oxford University Press, New York, 2001.

[21] GRIFFITHS, D.J., **Introduction to Quantum Mechanics**, Prentice-Hall, Upper Saddle River, NJ, 1995.

[22] HALLIDAY, D., RESNICK, R., AND WALKER, J., **Fundamentals of Physics**, John Wiley & Sons, Hoboken, NJ, Sixth Edition, 2001.

[23] HARTREE, D., *"The wave mechanics of an atom with non-Coulomb central field, Part 1: Theory and methods,"* **Proc. Comb. Phil. Soc.**, 24, pp. 89–312, 1928.

[24] HARTREE, D., **The Calculation of Atomic Structures**, John Wiley & Sons, New York, 1957.

[25] HARVEY, J., *"Molecular Electronic Structure— Lecture 4, The Hartree–Fock–Roothaan Method: Part 1,"* http://www.chm.bris.ac.uk/pt/harvey/elstruct/hf_method.html, 2001.

[26] HELLMANN, H., **Einführung in die Quantumchemic**, Deuticke, Liepzig, p. 285, 1937.

[27] HINCHLIFFE, A., **Computational Quantum Chemistry**, John Wiley & Sons, Chichester, 1988.

[28] HOHENBERG, P. AND KOHN, W., **Phys. Rev.**, 136, B864, 1964.

[29] JACKSON, J.D., **Classical Electrodynamics**, John Wiley & Sons, New York, Second Edition, 1962.

[30] KOHN, W. AND SHAM, L.J., *"Self-consistent equations including exchange and correlation effects,"* **Phys. Rev.**, 94, p. 111, 1954.

[31] KOHN, W. AND SHAM, L.J., **Phys. Rev.**, 140, A1133, 1965.

[32] KOHN, W., *"Nobel lecture: Electronic structure of matter–wave functions and density functionals"*, **Rev. Mod. Phys.**, 71(5), pp. 1253–1266, 1998.

[33] LEACH, A.R., **Molecular Modeling; Principles and Applications**, Pearson Education Limited, Prentice-Hall, Harlow, Second Edition, 2001.

[34] LIONS, P.L., *"Solutions of Hartree–Fock equations for Coulomb systems,"* **Commun. Math. Phys.**, 109, pp. 33–97, 1987.

[35] LIU, W.K., KARPOV, E.G., AND PARK, H.S., *"An introduction to computational nanomechanics and materials,"* **Computer Methods in Applied Mechanics and Engineering**, Vol. 193, pp. 1529–1578, 2004.

[36] MARDER, M.P., **Condensed Matter Physics**, Wiley Interscience, New York, 2000.

[37] MESSIAH, A., **Quantum Mechanics**, Volumes 1 and 2, Dover, Mineola, NY, 1999.

[38] PARR, R.G. AND YANG, W., **Density-Functional Theory of Atoms and Molecules**, Oxford University Press, New York, 1989.

[39] RATNER, M.A. AND SCHATZ, G.C., **Introduction to Quantum Mechanics in Chemistry**, Prentice-Hall, Upper Saddle River, 2001.

[40] SAAD, Y., CHELIKOWSKY, J.R., AND SHONTZ, S.M., *"Numerical methods for electronic structure calculations of materials,"* **SIAM Rev.**, Vol. 52, No. 1, pp. 3–54, 2010.

[41] SHERILL, C.O., *"An introduction to Hartree–Fock molecular orbital theory,"* http://vergil.chemistry.gatech.edu/notes/hf-intro/hf-intro.pdf, 2000.

[42] SLATER, V.C., AND KOSTER, G.F., *"Simplified LCAO method for periodic potential problem,"* **Phys. Rev.**, 94(6), pp. 1498–1524, 1954.

[43] VON NEUMANN, J., **Mathematical Foundations of Quantum Mechanics**, University Press, Princeton, NJ, 1955.

[44] ZETTILI, N., **Quantum Mechanics: Concepts and Applications**; Second Edition, John Wiley & Sons Ltd., Chichester, 2009.

Part III

[45] ALLEN, M.P., AND TILDESLEY, D.J., **Computer Simulation of Liquids**, Clarendan Press, Oxford, 1987.

[46] ANDREWS, F.C., **Equlibrium Statistical Mechanics**, John Wiley & Sons, New York, Second Edition, 1975.

[47] CERCIGNANI, C., **Rarefied Gas Dynamics: From Basic Concepts to Actual Calculations**, Cambridge University Press, Cambridge, 2000.

[48] CHANDLER, D., **Introduction to Modern Statistical Mechanics**, Oxford University Press, New York and Oxford, 1987.

[49] DILL, K.A. AND BROMBERG, S., **Molecular Driving Forces. Statistical Thermodynamics and Biology**, Garland Science, New York, 2003.

[50] EHRENFEST,P. AND EHRENFEST, T., *"Begriffliche Grunlagen der statistischen Auffassung in der Mechanik,"* Vol. IV, Part 2, **Enzyklopädie der Mathematischen Wissenschaften mit Einschluss ihrer Anwendungen**, Teubner, Leipzig, 1911.

[51] EKELAND, I. AND TEMAM, R., **Convex Analysis and Variational Principles**, North Holland, Amsterdam, 1976.

[52] FRENKEL, D. AND SMIT, B., **Understanding Molecular Simulation**, Academic Press, San Diego, Second Edition, 2002.

[53] GRANDY, W.T. JR., **Foundations of Statistical Mechanics: Vol. I, Equilibrium Theory**, D. Reidel, Dordrecht, 1987.

[54] GRANDY, W.T. JR., **Foundations of Statistical Mechanics: Vol. II, Nonequilibrium Phenomena**, D. Reidel, Dordrecht, 1988.

[55] HILL, T.L., **An Introduction to Statistical Thermodynamics**, Dover, New York, 1986.

[56] KHINCHIN, A.I., **Mathematical Foundations of Statistical Mechanics**, Dover, New York, 1949.

[57] LEACH, A.R., **Molecular Modeling; Principles and Applications**, Pearson Education Limited, Prentice-Hall, Harlow, Second Edition, 2001.

[58] LEBOWITZ, J.L., *"Hamiltonian Flows and Rigorous Results in Nonequilibrium Statistical Mechanics,"* in RICE, S.A., FREED, K.F. AND LIGHT, J.C. (eds.), **Statistical Mechanics: New Concepts, New Problems, New Applications**, Chicago Press, Chicago, 1972.

[59] McQUARRIE, D., **Statistical Mechanics**, University Science Books, Sausalito, CA, 2000.

[60] METROPOLIS, N., ROSENBLUTH, A.W., ROSENBLUTH, M.N., TELLER, A.H., AND TELLER, E., *"Equation of state calculations by fast computing machines,"* **J. Chem. Phys.**, 21, pp. 1087–1092, 1953.

[61] ODEN, J. T., **Qualitative Methods in Nonlinear Mechanics**, Prentice-Hall, Englewood Cliffs, NJ, 1986.

[62] PATHRIA, R.K., **Statistical Mechanics**, Butterworth-Heinemann, Oxford, Second Edition, 1996.

[63] PLANCHEREL, M., *"Beweis der Unmöglichkeit ergodischer mechanischer Systeme,"* **Ann. Phys.** 347, pp. 1061–1063, 1913.

[64] ROSENTHAL, A., *"Beweis der Unmöglichkeit ergodischer Gassysteme"*, **Ann. Phys.** 347, pp. 796–806, 1913.

[65] SCHROEDER, D.V., **An Introduction to Thermal Physics**, Addison-Wesley Longman, San Francisco, 2000.

[66] WEINER, J.H., **Statistical Mechanics of Elasticity**, Dover, New York, Second Edition, 2002.

INDEX

Ab initio method, 189

Absence of magnetic monopoles, 88

Absolute temperature, 50, 60, 233, 256, 257

Almansi–Hamel strain tensor, 8, 10

Amorphous solid, 180

Ampere, 76, 82

Ampere's law, 82, 83

Ampere–Maxwell Law, 83, 88

Angular frequency, 85, 86, 99, 132

Angular momentum, 27, 35, 36, 38, 39, 41, 52, 53, 84, 145, 146, 149–151, 154, 291, 307

   extrinsic, 145–148, 151, 152

   intrinsic, 84, 146, 156

Associated Laguerre polynomial, 137

Associated Legendre polynomial, 139

Assumption of molecular chaos, 267

Atomic bond, 165, 173

Atomic element, 165, 169

Atomic structure, 96, 135, 142, 161, 165, 173, 181, 243

Avogadro's number, 215

Azimuthal quantum number, 137, 151

Backbone of the molecule, 181

Balmer series, 138

Benzene, 176

Blackbody radiation, 95

Body force, 290

Bohr, 81, 108, 137, 138, 143, 174, 175

Bohr formula, 138

Bohr representation, 174

Boltzmann, Ludwig, 220, 233

   distribution, 237

      discrete, 236

   equation, 228, 233, 250, 265, 266, 268, 269

   factor, 264, 265

   H-theorem, 269

   inequality, 268, 269

Born, 104

Born–Oppenheimer approximation, 189, 190, 192, 194,
   195, 241
   surface, 194
Bosons, 156
Brackett series, 138
Bulk velocity, 270

Canonical ensemble, *see* Ensemble
Cauchy sequence, 117
Cauchy's theorem, 36, 37, 39, 289–
   291
   hypothesis, 32
   stress tensor, 36, 38, 39, 41, 59,
      60, 270, 289, 290, 292
Cell-biology molecules, 180
Characteristic function, 125
Characteristic function of a set, 222
Characteristic polynomial, 21
Charge density, 80, 83, 88, 202
Chemical bond, *see* Atomic bond
Clausius–Duhem inequality, 51,
   52, 56, 62
Closed, 49, 50, 82–84, 88, 224, 234
Cofactor, 7, 14, 283
Coleman–Noll method, 60–62
Collision invariant, 268–270
Collision term, 266
Commutator, 128, 151, 238, 304
Concept of stress, 30, 31
Configuration
   current, 4, 5, 7, 15, 18, 20, 25,
      26, 32, 33, 39, 59, 272,
      281, 287, 288
   definition, 3
   electronic, 168, 169, 174, 178

reference, 3–5, 7, 15, 18, 20,
   23, 26, 39–41, 43, 46, 47,
   53, 55, 58, 59, 61, 68, 281,
   287, 288, 291, 294
Configurational part, 263
Conservation of charge, 79, 87
Conserved, 76, 79, 266, 272
Constitutive equation, 53–61, 63,
   64, 66, 67, 69, 89, 293
Continua, 11, 49, 250
Continuous spectrum, 97, 125–
   127, 129
Controllable kinematic variables,
   252
Correspondence principle, 108,
   307
Coulomb, 75–77, 80, 82, 191, 198,
   199
Coulomb's Law, 75–77, 80
Covalent bond, 175–178, 244, 245
Crystalline solid, 180
Current, 82, 83, 88, 90, 95
   Ampere, 76
Current configuration, *see* Config-
   uration
Current flux density, 83

Deformable body, 3, 45, 49, 69,
   281, 287
   force and stress in, 29
   kinematics of, 3, 281
Deformation gradient, 6, 13, 19,
   68, 252, 282, 288
Deformation rate tensor, 14
Density functional theory, 189,
   200, 212

Density matrix, 199
Diffraction, 94, 97
Dimensional consistency, 57
Dirac, 82, 93, 103
    delta distribution, 230, 302
    equation, 103
Displacement, 6, 54, 89, 251, 282, 287, 293, 294
    gradient, 6
Distribution function, 105, 113, 215, 220–223, 225, 226, 229, 235–237, 239, 252, 253, 256, 258, 260, 265
DNA, 180, 185, 187
Dynamical variable, 115, 116, 118–121, 125–127, 146

eigenspin, 153
Eigenvalue problem, 21, 120, 121, 127, 193, 198
Electric charge, 75–77, 81, 82
Electric dipole, 78, 84
    moment, 79
Electric displacement field, 89
Electric field, 75, 77–79, 81–83, 87, 91, 92
Electromagnetic field theory, 73, 108
Electromagnetic waves, 75, 84, 91, 92, 94
Electron, 75, 76, 81, 82, 84, 92, 95–99, 103, 133, 134, 138, 139, 146, 150, 152–154, 156–161, 165, 167–171, 173–175, 177, 178, 189–192, 195–198, 200–202,

205–212, 241, 306, 307
    density, 161, 200, 201
Electronegativity, 177
Electronic structure, 154, 161, 165, 169, 170, 205, 212
Electrovalent bonding, 173
Energy
    conservation, *see* Principle of
    definition, 45, 250
    energy-wave relation, 95
    exchange-correlation, *see* Exchange-correlation energy
    flux, 270
    Gibbs free energy, *see* Gibbs free energy
    ground state, *see* Ground state energy
    Hartree–Fock, *see* Hartree–Fock method
    Helmholtz free energy, *see* Helmholtz free energy
    internal, 45, 46, 54, 60, 68, 218, 238, 239, 251, 254, 261, 271, 309, 311
    ionization, 138
    kinetic, 42, 43, 45, 46, 100, 191, 208–210, 218, 219, 238, 239, 250, 251
    potential, 70, 80, 102, 132, 219, 242, 243, 299
        general form of, 243
    shell, 81, 166, 170, 171
    spectrum, 95, 97
    strain, *see* Stored energy function

Engineering strain tensor, 10

Ensemble, 215, 220–225, 228, 229, 231, 234, 236, 240, 250, 314
    average, 107, 215, 220, 221, 236, 237, 239, 242, 252, 264, 265
    canonical, 224, 230, 234, 239, 252, 254, 262, 263, 310
    grand canonical, 224, 239, 240, 314, 315
    microcanonical, 223, 228–230, 232, 233

Entropy, 50, 51, 53, 54, 60, 66, 233, 257, 258, 261, 309, 311, 314
    flux, 51
    supply, 51

Equation of motion, 38, 39, 41, 218, 219, 224, 226, 227, 241, 245, 291

Equation of state, 66

Equipresence, 57

Ergodic hypothesis, 220, 221, 252

Ergodic system, 220–222
    quasi-, 221

Ethene, 176

Ethyne, 176, 179

Euler's theorem for first-order homogeneous functions, 314

Eulerian formulation, 11–13, 27

Exchange integral, 198, 199

Exchange-correlation energy, 208, 210, 212

Existence, 19, 57, 82, 97, 146, 180, 252, 289

Expected value, *see also* Mean, 107, 108, 113, 119–121, 125, 127, 206, 221, 237, 239, 299, 304

Exterior, 15, 29–32, 36, 49, 225, 286, 287, 291

Faraday's Law, 83, 88

Fermions, 156–158, 162, 197

First Hohenberg–Kohn theorem, 208

Fluid dynamics, 250

Fock, 194

Force
    body, 29–31, 36, 69, 72, 272, 292, 294
    concept of, 29
    conservative, 241
    contact, 29–32, 49, 69, 290, 294
        external, 30, 31
        internal, 30, 31
    deformable body, *see* Deformable body
    electrostatic, 75, 77, 80
    external, 29, 49, 243, 293
    generalized, 254, 260
        macroscopic, 253
        micromechanical, 253
    induced electromotive, *see* Faraday's Law
    internal, 251
    Lorentz, 82
    surface, 30

Fourier transform, 100, 101, 109, 110, 125, 127

Fourier's Law, 66, 68

Gas dynamics, 66, 250, 268
Gauss's Law, 79, 80, 84, 87
Gauss's Law for magnetic fields, 84
Generalized Laguerre polynomial, 137
Gibbs free energy, 262, 309, 313, 316
Gibbs relations, 250, 261, 262
Gibbs, Josiah Willard, 220, 221
Gibbs–Duhem equation, 314
Grand canonical ensemble, *see* Ensemble
Green–St. Venant strain tensor, 8, 10, 282, 288, 312
Ground state energy, 133, 141, 162, 163, 166, 189, 191, 194, 197, 206, 207, 212
Group velocity, 109
Gyromagnetic ratio, 154

Hamiltonian, 102–104, 121, 132–134, 140, 150, 154–157, 162, 163, 165, 190–192, 194–196, 206–208, 218–220, 224–227, 231, 234, 237, 239, 242, 252, 259, 260, 263, 304, 305
    conjugate, 259, 260
    electronic, 193–195
    nuclear, 193
    operator, 102, 104
    separable, 158
Hamiltonian mechanics, 216
Harmonic oscillator, 131–133

Hartree, 194
Hartree method, 194, 196, 198, 208
    potential, 205
    product, 196
    wave function, 196
Hartree–Fock method, 189, 194, 196, 197, 200, 208
    energy, 197, 199
    equations, 198, 199, 210, 212
Heat conduction, 67, 177
Heat equation, 68
Heisenberg, 97
    uncertainty principle, 97–99, 107, 128, 299, 302
    generalized, 127, 128
Helium atom, 157, 158, 163, 307
Helmholtz free energy, 58, 60, 61, 67, 258–261, 309–311, 313
Hermite polynomial, 133
Hermitian operator, 118–121, 124–127, 129, 130, 303
Hilbert space, 116, 117
    $H^1(\mathbb{R}^d)$, *see* Sobolev space
    $L^2(\mathbb{R}^d)$, 110, 116–121, 123–125, 127, 158
Hohenberg–Kohn theorem, 205, 208
Homogeneous, 6, 58, 69, 269, 313, 314
Hybridization, 178, 179
Hydrogen atom, 131, 133, 134, 136–138, 140, 147, 150, 151, 157, 158, 163, 173, 175–177, 179, 306
    orbitals, 140, 141, 165, 166

Hydrogen bond, 177, 178
Hyperelastic, 70

Ideal fluid, 59
Identical particles, 155–157, 263
Identities, 202, 256, 276, 277
Incompressible, 59, 60, 63, 64
Induced electric field, 83
Induced electromotive force, *see* Faraday's law
Inner product, 110, 116–118, 122, 209, 237, 279, 280, 284
Inner product space, 116, 117, 277, 278
Internal dissipation field, 62
Ion, 173
Ionic bond, 173, 178
Isochoric, 59
Isotope, 171
Isotropic, 56, 69, 71

Khinchin structure, 230, 231
Kinematics, definition, 4
Kinetic theory of gases, 228, 250, 266
Klein–Gordon equation, 103
Kohn–Sham theory, 208, 212
    equations, 211, 212
Kronecker delta, 277

Ladder property, 149
Lagrange multiplier, 20, 209–211, 235
Lagrangian dynamics, 218
Lagrangian formulation, 11, 12, 27
Lamé constant, 71

Left Cauchy–Green deformation tensor, 8, 19
Lennard-Jones potential, 243, 245
Line spectra, 138
Linear elasticity, 70
Linear momentum, 27, 35, 36, 38, 39, 41, 52, 53, 66, 145, 272
Linear transformation, 56, 277, 278
Liouville
    equation, 266
    operator, 228, 266
    theorem, 225–227
        corollary, 226, 227
Local density approximation, 212

Macroscopic balance laws, 269, 272
Magnetic dipole moment, *see* Spin
Magnetic field, 29, 81–84, 87–89, 91, 92, 154
Mass
    conservation, *see* Principle of definition, 25
    density, 25–27, 49, 267, 269, 270, 294
Material coordinate, 4, 6, 12, 14, 27, 56, 281
Material description, 11, 27, 51
Material time derivative, 13, 90
Maxwell distribution, 269
Maxwell's equations, 87, 89, 94
Mean, *see also* Expected value, 108, 113, 119, 120, 239
Mean-field approximation, 205
Metallic bond, 177, 178

Methane, 175, 176, 179

Metropolis method, 250, 262, 264, 265

Microcanonical distribution, 229

Microcanonical ensemble, *see* Ensemble

Microcanonical partition function, 229, 232, 233

Molecular dynamics, 5, 240, 241, 265

Molecular structure, 165, 175, 180, 184, 233

Momentum flux, 270

Momentum space, 110

Monopole, 81–83, 88

Monte Carlo method, 250, 262–264

Motion
    definition, 4
    orientation preserving, 5, 6

Multiplicity, 231, 232, 234, 240, 314

Nanson's formula, 18

Navier–Stokes equation, 64
    compressible, 66
        fluids, 66, 266, 269
        gases, 66, 266, 269
    incompressible, 63

Neutron, 75, 152, 171

Newtonian fluid, 60, 66

Non-Newtonian, 60

Nucleus, 75, 81, 133, 135, 146, 157, 158, 167, 168, 170, 171, 177, 190–195, 241, 306

Observable, 3, 112, 115, 119–121, 125–128, 149, 156, 237, 252, 301, 307

Particle in a box, 304

Partition function, 223, 229, 230, 232, 233, 257, 259–263, 310

Paschen series, 138

Pauli's principle, 145, 155–157, 160–162, 196, 197

Period (of oscillation), 85

Periodic table, 142, 161, 165, 169, 170, 173
    groups, 171

Permeability constant, 82

Permutation operator, 156, 157

Phase (of the wave), 86, 91, 92, 94, 99

Phase function, 219, 221, 229, 236, 237, 252–254, 258, 262

Phase space, 115, 217, 219, 222, 224, 225, 228, 229, 231, 242, 251, 262, 264, 267, 269

Phase velocity, 217, 270

Phosphodiester linkage, 181

Photon, 81, 92, 96, 97, 138, 152, 157

Piola transformation, 15, 18, 40

Piola–Kirchhoff stress tensor, 39, 293
    first, 40, 41
    second, 41, 43, 253, 312

Planck's energy quantum, 133

Planck's formula, 138

Poisson's ratio, 71

Polar bond, 177

Polar decomposition theorem, 18–20, 59

Potential, 105, 107, 132, 133, 191, 194–196, 202, 205–208, 211, 212, 241–243, 245, 250, 265, 292, 300, 302, 305, 306

  s-body, 243

  chemical, 210, 211, 224, 239, 309, 312

  electric, 80

  energy, *see* Energy

  energy functional (general form), *see* Energy

  interaction, 202

  Lennard-Jones, *see* Lennard-Jones potential

  thermodynamic, *see* Thermodynamics

Power, 42, 43, 45, 46, 251, 272, 312

Primary bond, 178

Principal direction, 20, 22, 287–290

Principal invariant, 20, 22, 23, 289

Principal quantum number, 137, 141, 166

Principal value, 22, 289, 290

Principle of

  balance of angular momentum, 35, 36, 38, 39, 52, 53, 291, 307

  balance of linear momentum, 35, 36, 39, 52, 53, 66, 272

  conservation of energy, 45–47, 52, 53, 56, 66, 68, 272

  conservation of mass, 26, 27, 52, 53, 56, 64, 271

  determinism, 54

  equal a priori probabilities, 228

  local action, 57

  material frame indifference, 55, 57, 59

  material symmetry, 56

  physical consistency, 56, 60

Probability density, 81, 99, 105, 111, 113, 115, 125, 141, 167, 200, 267, 298

Probability space, 111, 112

Proton, 75, 76, 134, 152, 157, 170, 171

Quantum mechanics, 73, 108, 110, 115, 121, 131, 133, 146, 155, 158, 194, 205, 232

Quantum model, 81

Quasi-static process, 249–252, 254, 255, 257, 258

Rate of deformation, 13, 54, 66

Rayleigh–Ritz method, 158, 161–163, 189, 194

Real vector space, 277

Reference configuration, *see* Configuration

Reflexive, 117

Relativistic quantum mechanics, 102, 103, 146

Resolution of the identity, 129

Response function, 54, 57, 58

Reynold's transport theorem, 23

Right Cauchy–Green deformation tensor, 7
Roothaan equations, 199, 200
Rutherford model, 80, 97
Rydberg's constant, 138
Rydberg's formula, 138

Schrödinger, 81
Schrödinger's equation, 93, 99, 101–105, 111, 115, 131–134, 150, 153, 158, 189, 194, 197, 234, 300, 303, 306
    electronic, 193
    for a free particle, 100
    in spherical coordinates, 135
    nuclear, 193
    time-independent, 104, 121, 135, 155, 191, 193, 300, 305, 306
Second Hohenberg–Kohn theorem, 208
Secondary bond, 178
Self-consistent field (SCF), 199
Separable, 117, 158, 306
Separation constant, 136, 138
Shear, 9, 66, 282, 287, 290
Shear strain, 9
Shear stress, 290
Shells of electron density, 161
Sigma bond, 175, 179
Simultaneous eigenfunctions, 130, 149, 301
Slater determinant, 160, 161, 197
Sobolev space, 117
    $H^1(\mathbb{R}^d)$, 116–118, 161

Spatial coordinate, 4, 27, 59, 115, 197, 281
Spatial description, 11, 12, 27, 316
Spectral theory, 121
Speed of light, 78, 82, 84, 91, 92, 97
Spherical harmonics, 139, 150, 151
Spin, 103, 145–147, 151–154, 156, 157, 159, 160, 197, 202, 208, 210
    angular momentum, 84, 154
    down, 152, 154, 168
    magnetic dipole moment, 84, 154
    moment, 155
    momentum, 168
    of the electron, 153, 154, 159, 161, 197
    operator, 152, 153, 307
    orbital, 153, 158, 160–162, 197, 199
    paired, 154, 161, 168
    Pauli's matrices, 153
    state, 153, 154, 159
    tensor, 14
    up, 152, 154, 168
Spinor, 103, 153, 154
Standard deviation, 98, 113, 298
Stationary, 109, 226
Statistical mechanics, 50, 215, 216, 219–221, 224, 231, 240, 249–252, 254, 255, 257, 263, 265, 268, 312
Stirling's formula, 235, 311
Stokes' theorem, 89

Stored (or strain) energy function, 69

Stress power, 42, 43, 312

Tensor product, 276, 278, 279
Tensor, definition, 278
Theory of elasticity, 69
Thermodynamics, 50, 215, 222, 223, 233, 249, 252
   equilibrium, 49, 223, 233, 239, 250
   first law, 46, 250, 251, 254, 255, 257, 312, 315
   potential, 223, 233, 240, 310
   process, 49, 251
   second law, 49, 51–53, 60, 233
   state, 49, 50, 53
   state variable, 49
   system, 49, 50, 262, 311
   temperature, 49, 233
Thermoelastic, 58, 295
Time average, 219–221, 241
Trace
   of an operator, 238
   of a tensor, 22, 280
   of a tensor product, 279
Trajectory, 5, 11, 145, 146, 193, 217, 219–222, 224–228
Trichloride, 177

Univalent, 173

Valence, 173, 175, 177
Van der Waals bond, 178, 244, 245
Variance, 113, 120, 128, 149, 298
Variational principle, 161, 162, 206
   method, 158, 208

Velocity gradient, 13, 58
Verlet algorithm, 245
Viscosity, 60
Voltage, 80
Vorticity, 14

Wave amplitude, 85, 101
Wave function components, 143
Wave function method, 189, 194, 200
Wave number, 85, 86, 93, 99, 138
Wave speed, 85–87, 92
Wave–momentum duality, 110
Wavelength, 85, 86, 91–97, 109, 138
Weak interaction, 231
Well-posedness, 57

Young's modulus, 71